THE THEORY
OF EVOLUTION

THE THEORY OF EVOLUTION

What It Is, Where It Came From, and Why It Works

CYNTHIA MILLS

WILEY

JOHN WILEY & SONS, INC.

Published by John Wiley & Sons, Inc., Hoboken, New Jersey
Published simultaneously in Canada

For general information about our other products and services, please contact our Customer Care Department within the United States at (800) 762-2974, outside the United States at (317) 572-3993 or fax (317) 572-4002.

Wiley also publishes its books in a variety of electronic formats. Some content that appears in print may not be available in electronic books. For more information about Wiley products, visit our web site at www.wiley.com.

ISBN 0-471-21484-1

Printed in the United States of America

10 9 8 7 6 5 4 3 2 1

Contents

Introduction   1

1 Making Up the Western Mind   5
2 Making Up Darwin's Mind   19
3 What Darwin Said   43
4 Reaction   69
5 Pea Plants, Flies, and the Modern Synthesis   97
6 Impact   131
7 Post-Normal Science   153

Glossary   191

Bibliography   205

Index   209

Introduction

How grand is the onward rush of science!
—Charles Darwin to Alfred Russel Wallace, 1872

Few ideas have engendered as much emotional resistance as the theory of evolution. Darwin's announcement generated voluminous and venomous reviews in newspapers, and turned many of his friends into enemies. Here in the twenty-first century, we find the situation little improved, with Christian government officials encouraging teachers in public schools to overlook evolution and Darwin entirely, as if the theory never did turn science entirely on its head. In response, the scientific community reacts with alarm and public ridicule.

Why is the theory of evolution so inflammatory? Why do so many nonscientists feel the need to valiantly attack or defend the idea? Theories like the Big Bang are as contradictory to religious beliefs, yet you seldom see Big Bang bumper stickers. I believe there are two reasons.

1

First, evolution is personal. It says something about each one of us—that we were once simpler creatures, that we didn't start out as the exalted and unique creations we now assume we are. In fact, we looked like apes. This can come as something of a shock, particularly if one is inclined to look down on our more hirsute cousins.

But another reason has to be that the theory of evolution, and in particular Darwin's theory of natural selection, is so accessible. It has the unmistakable elegance of a great theory; once you read it, you say, as Thomas Huxley did, "Why didn't I think of that?" Darwin's statement of his idea, *The Origin of Species,* is so clearly drawn that anyone can understand it, and understand the evidence he gives to support it. How many other scientific theories are like that? Most of us can't argue with Newtonian physics or Einstein's relativity with anything like expertise, because most of us can't reproduce the mathematics. Darwin himself struggled hard with mathematics. He had to abandon his attempts to quantify his ideas when his cousin Francis Galton pointed out several egregious miscalculations. The solitary graph in *Origin* is easy enough for anyone to understand.

Darwin's theory is so unmathematical that scientists still argue over whether or not it qualifies as a theory. Ernst Mayr, an unabashed Darwinian, describes *Origin* as "one long argument," comparing it to a lawyer's brief rather than a set of postulates and laws. Without a rigorously definite foundation, the theory takes on the air of a narrative—something less than scientific, something assailable.

That very vulnerability encouraged many to attack. Even those who agreed with Darwin felt confident enough to quibble over details. Maybe they knew Darwin was right about the fact that species change, but they could disagree with natural selection, his idea of how they changed. If they agreed with selection, they rejected his claims that change was gradual. When scientists disagreed with Darwin, they opened the way for nonscientists. If fault could be found among Darwin supporters, the religious detractors felt the theory was a sitting duck.

But the greatest damage to the Darwinian reputation was done by those who would co-opt evolution and natural selection for their own purposes. Darwin's theory was quickly adopted, misinterpreted, and rewritten to promote various unsavory ideologies. No one, certainly not Darwin, would argue there is any real value to social Darwinism. And we need not credit those who would support the role evolutionary theory played in the justification of the Nazi Holocaust.

The saddest yield of all this opposition is the misrepresentation of Darwin himself. He was a product of his times, certainly, and undoubtedly held imperialist and racist views when held to twenty-first-century standards of propriety. At the same time, he was profoundly antislavery, and he couldn't bring himself to entirely agree with his cousin Francis Galton on eugenics. Mainly, he was far from arrogant or egotistical; he was ambitious enough to know he had a good idea and cunning enough to set the stage for it. Still, he clung desperately to his Victorian sense of integrity and honor. He avoided the limelight, but I think he had something he desperately wanted to say about the world. I think he wanted to put us back into nature.

Darwin had a bad case of what E. O. Wilson calls "biophilia." He couldn't look at nature without being absorbed by it, and he couldn't bring himself to feel above or separate from the creatures he saw around him. He saw no big jumps between us and the rest of animalkind; all of nature is made of the same stuff. The difference between a human mind and that of an orangutan was one of degree, not of kind.

This book is not a biography of Darwin, however. And Darwin probably was wrong about much of what he believed. The theory of evolution, like all theories, is larger than its originator. To meet the criteria of a great idea it must take on a life of its own, and evolution has certainly done that. It has built a science, several in fact, creating its own turmoil, without and within. Within biology, factions form to argue over whether species change suddenly or gradually. Other factions argue whether selection is the sole driving force, while others insist on giving it a minor walk-on part. And then there are the outlyers, the creationists and intelligent-design supporters, always looking for a weak point to attack.

The following is a biography of this contentious theory, one that describes its genealogy, birth, growth, education, and its impact; perhaps not its death. Like all scientific theories, it had a long and gradual birth, and a great many premature deliveries. There were a lot of good ideas that laid the groundwork, and many of these same good ideas became bad ones that bogged down progress. But what made evolution such a good idea? How has it withstood critics in greater number and of greater passion than any other scientific concept?

4

1

Making Up the Western Mind

Once upon a time we humans looked out over a world of nature that didn't change. What we saw with our own eyes was what we got. A duck was a duck because it was supposed to be, because there was some essential nature—an essence—that made it that way.

There was room for argument. Plato and his student Aristotle disagreed over Plato's idea that this essence was preset—as if there were some great ideal model for ducks set by some greater intelligence. Aristotle looked at each creature and decided the essence of the duck came from the duck itself, as if the duck came with the instructions for its own creation. Still, a duck was a duck.

The world didn't change, but it didn't sit still, either. If people watched the skies, they saw that the sun moved, and the planets, and so did the stars. And old things died, and new things were born. But all this story repeated itself over and over. People decided the essence of the world was circular. Time ran in cycles, coming back again and again to the same place. The heavenly bodies moved, but the earth didn't.

Then came the Dark Ages, and the study of nature took a back seat to religion, in particular to the Christian religion. While scholars debated the details of what Jesus said, they mostly accepted the writings the Greeks left behind about nature. They differed in one important way: with Christianity came a Bible, and the Bible told a story with a beginning and the prophesy of an end. There was no coming back to square one in this book. The Christian world had a future and a past, having more or less invented the concepts.

For a long time this disturbed no one. Indeed, people didn't seem to notice the difference, and didn't change their opinions of nature, perhaps because only a few people thought about nature at all. The others were busy struggling to stay alive, and didn't read or write much. The few scholars that did concentrated on the glory of God, for the most part. Nature, if they considered it at all, was merely appreciated as an example of God's creative genius.

Then the world started to open up. People started to look around. Civilization advanced enough to give some people the spare time to go exploring. And other people began to learn to read and write; more people gained the wherewithal to build their own opinions.

There were the famous travelers. Marco Polo had gone east and returned; Christopher Columbus had gone west.

They brought stories and samples of new peoples, new plants, and new animals. Unimaginable creatures were brought home to gawk at, collect, and study.

These discoveries demonstrated two things: there was far more life on earth, and much more variety, than expected; and each and every species was remarkably well adapted to its way of life. There were so many new creatures, they began to overcrowd the Ark. It became uncomfortably clear that there wouldn't have been enough room on Noah's homebuilt boat for all the world's creatures. Christian scholars could lighten the load by tossing off the birds and marine organisms, but even that wasn't enough. There was another problem. How could creatures so perfectly adapted to one environment, say the polar bear or the rain forest sloth, survive a long migration back home starting in the desert regions of Mt. Ararat?

There was also the discovery of a whole new world. With the invention of the microscope, Anton van Leeuwenhoek's tiny "animalcules" extended biology into an unheard-of realm. Scripture didn't even mention such creation—how could they now be accounted for? The existence of such microscopic creatures also gave support, if false, to the theory of spontaneous generation—the idea that living forms can spontaneously arise from nonliving material. The theory had lost some of its hold after some demonstrations that flies would not arise out of rotten meat if flies were not allowed to lay eggs. But a handful of hay left in water still resulted in the generation a few days later of hundreds of tiny, busy, microscopic creatures. Where had they come from? Though spontaneous generation was later disproven, it was a theory for the creation of life that didn't require God's hand. The discovery of microscopic forms was yet another chip in the foundation stone of sacred texts.

Yet another world emerged, a world out of time. The ancient Greeks described fossils, and Aristotle, finding their forms consistent with animal forms, likened them to animals. Still, little significance had been attached to their existence. The Greeks explained them as organic forms, arising from some sort of abstract accident. During the Enlightenment, theories linked fossils to spontaneous generation—it was suggested that fossils were the end result of misplaced germs, attempting to come to life in apparently inappropriate surroundings. Still, as it became clear that the visible surface of the earth had a decipherable past, and fossils were parts of that past, more and more scientists collected, analyzed, and described different fossils. It was becoming harder to deny their complexity, and to disavow what they said about the world.

Instead of the old theist worldview, where God takes a direct hand in every creation, God was taking a step back, becoming the watchmaker supreme: the designer of the *forces* that made our universe run. He no longer had to infuse energy and matter with his genius on an everyday basis; he just had to wind it up and let it go. This was the new framework, a perfect mechanism that could be taken apart and explained. And every step of this process brought the investigators a step closer to reading the mind of God.

At the turn of the eighteenth century, Paris was becoming the hub of this new rational biology. Explorations around the world were sending large collections of exotic plant and animal specimens back to the city, and museums had to be built to house them. The Jardin du Roi, later renamed the Museum of Natural History (or Museum du Histoire Naturelle) became a hotbed of innovative thought and argument. Buffon, Cuvier, Lamarck, and Geoffrey St. Hilaire performed research and debated their ideas there.

BUFFON

Few scientists have represented their time as fully as Georges Louis Leclerc, comte de Buffon. Buffon, as he is commonly known, was born in 1707, the son of a wealthy aristocratic family. He studied one year in England, where he developed a passion for physics and mathematics, and also studied plant physiology. He was especially inspired by reading the works of Isaac Newton, even translating Newton's *Fluxions* into French upon his return.

Although not primarily a biologist, he was recommended for the directorship of the Jardin du Roi. He took to the job with characteristic flair and not a little bluster. He decided his role was to Newtonize biology and set about compiling an encyclopedia of natural history. Buffon wrote thirty-five large volumes, called *Histoire Naturelle*, between 1749 and 1788.

These volumes reflected not only Buffon's ideas (which changed in progress, sometimes radically), they also reflected the political and scientific pressures of the times. Initially, flush with Newtonian fervor and inspired by such social writers as Diderot, Buffon in his first volumes denied the existence of species at all, claiming there were only individuals. In part he was joining the philosophers in defying religion on behalf of scholarship. He was also trying to draw a picture of biology analogous to Newton's view of the physical world, based on mechanistic principles and not what he saw as obsolete abstractions.

He was also reacting to a trend of the day, the classification of species by a few arbitrary characteristics. In the same vein as Aristotle, Buffon was turning back to observational science, gaining scientific understanding of species by cataloguing specifically what he saw. He was particularly critical of his contemporary, the Swedish botanist Carolus Linnaeus. Linnaeus in particular clung to an essentialist approach—a system declaring that elements of nature, including species, had an ideal and constant essence or quality that did not change. He built his taxonomy based on the presence or absence of selected, predetermined characteristics. Buffon called biologists like Linnauus *nomenclateurs,* insinuating they did nothing but give names.

The possibility of species change, or mutability, reared its ugly head about this time. The realization that fossils represented animals that no longer existed challenged the preconceived ideas of the permanent nature of species. Linnaeus considered the possibility that species were not permanent and unchanging, but rejected it. He considered hybridization as a possible source of new species. Though wrong in this instance, Linnaeus did contribute a great deal to the theory of evolution unintentionally, through his classification system. Linnaeus was the first to group organisms more or less in parallel, rather than along a single, linear scale of progression. He scrapped any concept of the *scala naturae*—Aristotle's organization of nature in a hierarchical, linear fashion, from the least perfect atom to the epitome of perfection: man. Linnaeus's animal kingdom splayed out into a treelike structure, with species families of equal, if differing, complexity at the ends of branches.

Despite his early bluster, Buffon's later volumes brought species back into the picture—with no apologies and no explanations. He simply could not deny the convenience such classifying "nomenclature" provided. At the same time, he was doing seminal work. He began a new style of comparative anatomy that could detect unity of type across many species; any differences indicated a divergence from type. He conceived of what he called a *moule interiour,* or internal mold, which informed each embryo in its development. Again, like Aristotle, he anticipated genetic coding.

With specimens coming in from explorers around the world, he began to notice similarities in different species from similar climates. He proposed that the environment could have an effect on species, with similar environments changing animals in similar ways. That was how he (and others) explained variation in domesticated animals: they changed to adapt to each climate they arrived in.

This explanation set up a problem. By finding that animals are built on a common plan with modifications, suddenly he could not squelch an idea: Species came from other species. He quickly rejected the idea. In some of his writings it appears that the rejection arose from outside—he was subject to the powerful influence of the leading academy, the Sorbonne, which remained a strongly religious institution. The conflict is apparent in his writing:

> If it were admitted that the ass is of the family of the horse, and different from the horse only because it has varied from the original form, one could equally well say that the ape is of the family of man, that he is a degenerate man, that man and ape have a common origin; that, in fact, all the families, among plants as well as animals have come from a single stock, and that all animals are descended from a single animal, from which has sprung in the course of time, as a result of progress or of degeneration, all the other races of animals . . . but this is by no means a proper representation of nature. We are assured by the authority of revelation that all animals have participated equally in the grace of direct creation.

Much of Buffon's evolution is degeneration, or de-evolution. Since God created a perfect world, if anything changes it is a fall from grace. The ass and the ape were once perfect—they started as horse and man—but these creatures changed, and not for the better. This was nascent evolution, change for the worse, to some degenerate form.

But even when challenged, Buffon could not shake the essentialist view. He was like someone peering over the end of the high diving board, unable to bring himself to jump. Species were species; they might change, but only into varieties after all. In the end he firmly rejected the idea that an animal can change into a completely different species, giving three reasons. The first took the long view: during re-corded history not a single species had changed into a new species. Not only was there no witnessed change, there was no evidence of change, or some "intermediate" entity. Since Buffon believed in plenitude—the notion that everything that could exist does—there should have been a multitude of species representing the gradual change of one form to another, and there wasn't. Not only were there no living intermediaries, there was a paucity of fossil intermediaries as well. Finally, there was evidence that change would be unsuccessful: hybrids, when they did occur, were sterile.

Ironically, as he stifled the idea, he also gave it impetus. Simply by arguing against evolution he gave it a place at the table. Many of the questions he asked were to be answered, finally, only by evolutionary theory.

LAMARCK

The first person to finally take the leap and assert that spe-cies can change was a protege of Buffon, Jean Baptiste Pierre Antoine de Monet, Chevalier de Lamarck. He was the eleventh child born to a noble but poor family with a strong military tradition. Lamarck served in the army and fought in the Seven Years War, staying on as a soldier when peace was declared but finally resigning due to injury. He was left to eke out a living on his tiny pension in Paris. He supple-mented his income by writing, at first just for dictionaries. He developed an intense interest in botany, finally writing a four-volume flora of France. This publication brought him to Buffon's attention, and he hired Lamarck as a traveling companion and tutor to his son. Buffon also obtained for him a poorly paid position as assistant in the botanical department of the Museum du Histoire Naturelle. He worked there for five years, beginning in 1788.

Buffon died that same year, and without his sponsor Lamarck lost standing at work. In 1793 (the same year Louis XVI and Marie Antoinette went to the guillotine), after a reorganization of the museum undertaken at Lamarck's suggestion, he was assigned a professorship of the inferior animals, worms and insects. These were considered of little interest and afforded the professor of this group little status and even less glory. After all, these animals had received only cursory examination by the likes of Linnaeus, who lumped them together and called them all, simply, worms. Lamarck, ever the optimist, took to the job and essentially created a whole new field of biology. He renamed the group invertebrates, and brilliantly reorganized the constituents into the classification scheme we currently use. And he asserted the importance of their study:

> We should chiefly devote our attention to the invertebrate animals, because their enormous multiplicity in nature, the singular diversity of their systems of organization, and of their means of multiplication . . . show us, much better than the higher animals, the true course of nature, and the means which she has used and which she still unceasingly employs to give existence to all the living bodies of which we have knowledge.

For seven years he gave the course on invertebrates, say-ing essentially the same things year after year. Then, sud-denly, at fifty-five, he underwent a conversion. In 1800 his lectures completely changed, revealing an entirely new approach to the natural world, one in which species *could* change, and did.

What happened? Lamarck had already been impressed by the tendency of organisms to progress from simple forms to the more complex. He replaced the traditional chain of existence, a progressing toward perfection (i.e., man), with one progressing more generally toward complexity. Given this disposition, it took one more thing to convert him completely.

A friend at the museum, Jean Guillaume Bruguiere, died, leaving his collection of mollusks—living specimens as well as fossils—to Lamarck. As Lamarck examined and orga-nized this collection, he began to see a pattern. Grouping species together, he found he could group fossil species with living species; he found some fossil species to be anal-ogous to living species. He could create chronological, lin-ear sequences of these groupings, tracing a gradual devel-opment from the older species to the new over the ages.

This must have come as a tremendous relief to Lamarck. He did not, as did many of his contemporaries, deny the existence of a creator, often citing a "supreme author of all things," or referring to "His infinite power." With this essen-tialist and creationist viewpoint, Lamarck found the possi-bility of extinction hard to accept. For, if some species died out and vanished, why were there still so many, unless new ones were still being created? This would require occasional interference from the creator, which, as a deist, Lamarck couldn't accept. Now Lamarck had the perfect counter explanation: old species didn't die out—they just became new species.

Speciation—the changing of species—solved even more problems for Lamarck. A student of geology, he believed in a changing earth. If the earth changed, species would be continually out of sync, losing their remarkable adaptedness, unless they changed along with their surroundings. Lamarck believed these changes were initiated by need. As an organism faced new challenges it would have to struggle and strain to attain what it needed. The result of this struggle, Lamarck asserted, was a changed organism. Lamarck's classic example was the giraffe. As conditions made trees grow farther apart and taller, the original giraffes had to stretch their necks more and more to get to leaves. "Nervous fluid," which he and his contemporaries believed circulated through the nerves creating sensation and motion, would then be drawn to their necks, resulting in more growth. As the fluid went to one area, or organ, it would decrease in another, unused area, resulting in atrophy.

Then, Lamarck postulated in his second law, the changes would be passed on to offspring. Lamarck insisted the changes would be very slow and gradual. But there was no need for anyone to assert that a given species became extinct—it had merely met new needs by becoming an entirely new species with an entirely new, well-adapted form.

Lamarck's theory tends to inspire scorn today. Part of the reason is that it's quite wrong. But a good part of it was due to how his ideas were used and abused. Social thinkers found in Lamarck the perfect supporting ideas for their own agendas. Organisms didn't just change; they, like self-improving humans, willed the change. They became more perfect because they wanted to. This did not reflect Lamarck's ideas, and left him open to scorn—what giraffe had the presence of mind to "decide" to lengthen its neck? Add to this the general distaste for the results of the French Enlightenment—the Revolution and subsequent Terror—and you have a recipe for distrust and distaste for Lamarckian ideas that persisted for centuries. Yet he broke new ground, and was tremendously bold in claiming that species could and did change and in providing a mechanism—however wrong—for that change.

2

Making Up Darwin's Mind

By the time Lamarck died and Charles Darwin was born, Western thinkers had broached the idea of an inconstant natural world. They'd mostly wrestled it down, but not without effort. Essentialism was fading, but what it left in its wake was confusion.

The world expanded in all directions, even into other realms. There were new continents, new worlds of organisms too small to have ever been imagined. The earth itself showed off its ancient side, with fossils and sediments stretching history back longer and longer. Monarchies were torn down and replaced with more egalitarian constitutional governments. These changes tested old explanations and found them wanting. Those who were loyal to old ideas found it harder and harder make the pieces fit.

And people were seeing themselves able to understand—even influence—the natural order. Human minds could reduce gravity and momentum to equations, and chemical reactions to the combining of elemental particles called atoms. This success led to heady optimism: maybe everything could be explained on these terms and a materialistic world imagined. If the mind were capable of such feats, maybe humans even had the capacity to explain, even improve themselves.

One characteristic of nature still filled naturalists with something like religious awe. Living creatures were so well suited to their place; as anatomists dissected and naturalists observed, animals and plants proved themselves again and again to be exquisitely adapted. Such perfection could only reflect glory on their creator. To study life was to peek into the mind of God and honor him. Religious fervor drove the study of nature, yet at the same time the discoveries being made could not help but foment insurrection. The closer the scientists looked, the more they saw that contradicted their religious texts.

In some ways, Charles Darwin couldn't help becoming a revolutionary, despite how strongly he resisted. He was born into it—it was a family tradition.

His grandfather, Erasmus Darwin, was a physician who lived at about the same time as Lamarck. The two came to the same conclusions about species, or very nearly, although they neither met nor knew of each other. Erasmus was inspired by the writings of the French Rationalists, and was eager for the spread of "the happy contagion of liberty." He had nothing but scorn for the powers that were, saying that "a goose may govern a kingdom" as well as any "idiot . . .

in his royal senses." He rejected the tight, irrational bonds of the English monarchy, which ruled hand in hand with the Anglican church. Although not an atheist, he did question the authority of the Bible and the ascendency of Christ. He chose instead to revere science as his way of honoring creation.

Darwin's grandfather was a very successful physician. He dabbled in other areas of science and invention, creating a steering mechanism for carriages and a speaking-machine. His work made him wealthy and his fortune was expanded by a second marriage to the illegitimate daughter of an earl. (He fell in love with this woman while she was still married to her rich husband, who shortly died.) He had twelve children by two wives and two children by a governess. As may be surmised, he was quite a sensualist and wrote erotic poetry, as well as prescribing sex as a treatment for patients suffering from hypochondria.

Besides eroticism, he also wrote about science, perhaps more for the poetry than for the theory. He wrote a book, *Zoonomia,* all in verse, which described, among other things, the origin of life:

> Nurs'd by warm sun-beams in primeval caves
> Organic Life began beneath the waves . . .
> Hence without parent by spontaneous birth
> Rise the first specks of animated earth.

Erasmus was also great friends with a neighbor, Josiah Wedgwood, who was to be Charles Darwin's other grandfather. Wedgwood started life as a potter and launched a fine ceramics dynasty that survives to this day. He was a self-made man and, like Erasmus Darwin, believed that people should be free to make their own success. He extended this freedom to include religion, as well, although not quite so radically as Erasmus. He became a Unitarian, admiring the minister Joseph Priestley. Priestley taught that God ordained happiness for everyone in this world, not simply the noble and wealthy. The discipline also emphasized the physical world, discarding the miracles and mysteries of the Anglican church.

They met via an organization called the Lunar Society, called themselves "Lunatiks," and revered the new industrial, technical world. They enjoyed this new culture at least until shortly after the French Revolution and subsequent Terror. The horrors of this time filled the British government and public with disgust, feelings that extended to the ideals founding these movements. Riots gutted Priestley's chapel and drove him to America. Facing government sanction and oppression, Erasmus Darwin and Josiah Wedgwood's days of democracy and libertine behavior were over. They toned down their rhetoric, and Dr. Darwin gave up his hopes of becoming England's Poet Laureate.

Charles Darwin knew of his grandfather's fame, and read *Zoonomia*. Still, given the repression of the times, it is not clear that he took it seriously. Darwin's father, Robert, was also a physician, but less ambitious and not so scientifically inclined. He was also a large and strict man, who, after the death of his wife, tended to run an oppressive household. Charles was raised by his three older sisters, who also tended to discourage his early love of wandering the woods and collecting insects—his sister Caroline refused to let him kill or keep them, saying it was cruel.

His father sent Charles and his older brother to an Anglican boarding school, where the emphasis was on Latin and scripture, subjects stultifyingly boring to Darwin. He preferred exploring the mysteries of chemistry with his brother in their shed, earning the nickname Gas Darwin. His instructors considered this a trifling pastime. Later, when he was old enough, his devoted free time to riding and hunting. This propensity for such trivial pursuits alarmed his father, who thought Charles likely to fritter away his life like some wealthy nobleman. To forestall this, he sent his son off to medical school as soon as he could; Darwin arrived in Edinburgh at the tender age of sixteen.

At that time Scotland was far enough away from London to remain a sort of northern extension of the Enlightenment. Learning was a popular pastime in the city; lectures were everywhere, and the most entertaining instructors drew large crowds. Many of the lectures offered the most recent ideas in geology and zoology.

Still, Edinburgh was icy cold and Darwin hated most of medicine. One anatomy professor appalled him with his filthy habits and lectures. He stopped attending surgery demonstrations—they must have been harrowing affairs, with patients strapped down without anesthesia—after running from one performed on a child. He was also too young and undisciplined for the study. Besides, he was beginning to suspect he would inherit enough money to not have to worry about a profession.

Although his father undoubtedly disagreed, Darwin's time was not entirely misspent. He learned taxidermy from a freed slave from Guiana, South America, and listened to his stories of the tropics. And he found a mentor in one professor, Robert Grant, a biologist who had studied in France. Grant was in the process of becoming a world authority on invertebrates, with a special interest in their larval forms. He was also a devotee of Lamarck. He liked Darwin, since he was the grandson of Erasmus, whose *Zoonomia* he had cited in his thesis. So Darwin was exposed to the controversial idea of evolution. He was also a member of the Plinian society, a student group for biologists with radical ideas questioning creation and the separation of soul and body. Some students were overwhelmed by the intellectual and spiritual challenges of these new ideas. The intellectual excitement affected Darwin, too. But he remained guarded in his enthusiasm, caution bred into him through the experiences of his grandfathers on how society could view such unconventional thinking. Besides, working with Grant he was entering scientific society, even earning his own small publication on discovering the free-living, mobile state of sea mat larvae.

Scotland was also a center for the study and development of moral philosophy—a discipline later divided into philosophy, sociology, and economics. Although Darwin didn't study these subjects, the ideas were pervasive. Many of the new ideas fit with the ideals of his grandfathers. Adam Smith in Glasgow had detailed the laws of the marketplace in his *Wealth of Nations*. He described how individuals, while seeing to their own benefit, worked for the good of society. Each person in the market, by figuring out for himself what he could sell and how much, and how much it would cost to produce, ultimately found the most efficient production schedule. If, for example, there was too much competition in making bread, the individual might turn to making pasta. The result would be diversification and distribution of labor and effort that would serve everyone in the end. Darwin would later apply this concept of individual competition and diversification to biology.

For all these thinkers, no greater role model existed than Sir Isaac Newton. Newton was a scientific superstar, having established, with his book *Principia,* the ultimate set of scientific standards. He showed the world how nature should be described, writing a set of laws that could be demonstrated, clearly and mathematically, over and over again. Everyone who followed had to try to fit their theory, their model—no matter what it described—into some equally rigorous form. Social scientists such as David Hume, another Edinburgh resident, Adam Smith, and Thomas Malthus were looking to create their own, equally rigorous models.

NATURALIST CLERGYMAN

Darwin's success in natural history did not appease his father. Exasperated with Charles's failure in medicine, he laid down the law: no money, no support until he gained a profession. Dr. Robert Darwin had found the perfect profession for his squeamish son. A job with good financial rewards, lots of free time to pursue his invertebrate collecting hobby, and his own place in the country. He was to study in Cambridge to join the clergy.

But first Darwin had to think about it. After all, he'd just spent a year in the company of friends and mentors who'd challenged religion and found it wanting. On the other hand, Charles could read between the lines. He gave it some thought and read a bit to shore up his wavering faith. He headed to Cambridge.

This meant a shift back to reading boring classics. It also meant entering a community with an entirely different take on scholarship. In contrast to Edinburgh's open lectures, catering to the whole community, suddenly he was in a city where students lived a life separate from "ordinary" residents, where being caught in town and out of your academic gown could lead to "rustication," or expulsion. Some professors also served as policemen, or proctors, whose chief duty was the rousting of ladies of doubtful repute from the school community.

In lieu of ladies, there was eating, drinking, and gambling. Darwin spent a term or so engaging a bit too enthusiastically in these pastimes, to the diminishment of his father's pocketbook. But he wasn't too estranged from his former passions—he met his cousin William Darwin Fox at Cambridge, and joined him in the current craze of beetling.

Collecting beetles and comparing collections was at the time a nationwide fad. It fit in with religious devotion, for the variety and cleverness of God could be glorified by admiring the variety and multitude of these shiny insects.

Much, in fact, of Darwin's religious studies were devoted to the subject known as "natural theology." Promoted in particular by William Paley, this was the study of the natural world as the ultimate example and evidence for God's creation. The clearly stated logic of Paley's book appealed to Darwin, and helped him fit himself into the image of becoming a clergyman.

So, too, did other events. During his time at Cambridge, two anti-Christian radicals came to speak. The response of the establishment clergy was swift, ruthless, and vengeful. Darwin was to remember and fear for most of his life the ostracism these two men experienced.

Darwin became close friends with a Cambridge don, the Reverend John Henslow. Henslow, the sort of clergyman-scientist Darwin could imagine himself being, was supremely competent in all areas of natural science. He set up field trips for the students to wander the hillsides to collect insects, rocks, and plants. Darwin reveled in studying with him and gained a fond supporter in Henslow. He became Henslow's favorite student and became known to the other dons as "the man who walks with Henslow."

Henslow was to be the man who set Darwin on his path to glory. He gave Darwin the works of Sir John Herschel on the role of science, and Alexander von Humboldt's account of his journey to South America. In a sense, he prepared Darwin for the life he almost choose for himself, one of travel and adventure in the service of science. To further prepare Darwin, he decided to train him in geology, not by himself but by introducing him to Adam Sedgwick. Darwin stood in for Henslow to be Sedgwick's assistant on a survey

in North Wales. Charles spent a summer collecting and surveying the red sandstone of Wales, learning to be an effective geologist.

Geology was a revolutionary science in those days.
There were strong arguments over the age of the earth and
its history. While Bishop James Ussher carefully analyzed
medieval history to find the date of the creation of the
earth (9:00 A.M. on October 23, 4004 B.C.), biologists such
as Cuvier and Buffon gave a modern alternative. Their
analysis of rocks and fossils made it hard to believe the
earth was any less than several thousands of years old. And
they offered a metaphorical alternative to the six days in
the form of six epochs. But how did the earth come to look
the way it did? And how did the fossils end up where they
did?

At first there were two competing theories, presenting
opposing forces for formation. The first claimed the earth
arose out of the oceans; this theory was called neptunism, a
theory Darwin had learned in Edinburgh. The second, vulcanism, described the earth's history as one of cooling from
molten lava. Both tended to emphasize a tumultuous past,
with mountains and rifts coming about as the result of
tremendous cataclysmic forces—events like Noah's flood.
These theories were classified as catastrophism, meaning
they emphasized large, catastrophic events as responsible
for changing the earth.

Filled with the scientific and traveling fervor of Herschel and Humboldt, Darwin arranged and planned a trip
to the Canary Islands. He convinced Henslow and a few

others to go along, and his father agreed to front the cost. The trip did not come off, however; with the birth of Henslow's daughter and the death of another companion, plans fell through.

Darwin had little time to brood over the lost expedition. Returning home from Wales he found an alternative proposal: a voyage around the world on a surveying vessel, the HMS *Beagle*. The captain, Robert Fitzroy, facing his first command, wanted to hire a companion—ideally a naturalist. The ship was to travel around the world for two years, charting the shores of South America, finding and establishing inroads there for the British Empire.

The position was offered to Henslow, but with his wife and new child to care for, he couldn't accept. He knew the experience would be great for Darwin; he himself had longed for a similar adventure at the same age. He knew Charles was an enthusiastic, if unfinished, naturalist, and such a journey would suit him as he would it.

Charles was beside himself with excitement. He was dying to go. But he had to ask his father. Robert Darwin was quite sure the trip was a waste of time and clearly dangerous. Fortunately for Charles, his uncle, Josiah Wedgwood, thought it would be the perfect thing for Charles to do, and persuaded his father it would be "character building." Robert Darwin's mind was changed, and he not only was willing to allow Charles to go, but agreed to put up the 500 pounds to pay for it.

It was not a paid position. The *Beagle* already had a paid naturalist, Robert McCormick, the ship's surgeon, who would also catalogue the various species seen on the trip. Ship's surgeons in those days were not on the same social level as officers, however, and Fitzroy did not consider McCormick a suitable companion. Fitzroy was the scion of a longtime military and noble family. The *Beagle*'s first captain had shot himself off the South American coast, and with an uncle who'd slit his own throat, Fitzroy, anticipating the loneliness of such an expedition, feared for his own sanity. Officers in those times would have nothing to do with their men—they considered it a breach of discipline—and thus he faced the whole journey alone. Besides, Fitzroy was quite sure he wouldn't enjoy the company of anyone but a proper gentleman anyway.

Hearing that Henslow could not go, he at first rejected the suggestion of Darwin. His standards were so high that, given Darwin's middle-class geneaology, he was doubtful Darwin would qualify as a gentleman's companion. Beyond his merchant background, the taint of his rabble-rousing, free-thinking grandfather, Erasmus, may have made Charles even less attractive. Only as the departure day grew nearer and no other alternatives appeared did he agree to a trial meeting with Darwin.

He was pleasantly surprised. Darwin was nothing if not agreeable, supremely friendly, and suitably conscious of gentlemanly social conventions. Darwin was, after all, not a recent entry into the moneyed class—there were two generations before him. The Darwins and Wedgwoods were nearly the elite class.

So the deal was made. Charles was to get his adventure.

THE HMS BEAGLE

If Charles Darwin had not made his journey on the *Beagle,* we would not speak of Darwinism. Darwin was well on his way to a village parish, insect collecting and preaching, but this tour changed all that. Over the next five years (the journey stretched on and on), he was to see mind-opening things that would change his outlook forever. He would live through an earthquake. He would find fossils of as-yet-unheard-of creatures. He was to meet primitive tribes whose lives would appall him, and other tribes who would impress him with their civility and honor. His experiences strengthened his ideas and forged them into convictions. His discoveries would shore him up and make him certain of the value of his own ideas. He had been an intellectual hero worshiper; his voyage on the *Beagle* made it possible for him to imagine himself Newton's equal.

Most of these were not self-conscious changes. Darwin was ready for change but not aware of it. He instead concentrated on enjoying his experience, which wasn't always easy. For one thing, he had to live with Captain Fitzroy.

If there was one thing Darwin had learned about life from his childhood, it was how to be discreet. While his grandparents cherished a free-thinking, Whiggish political outlook, they kept it quiet. So could Darwin. He could go to Anglican schools while hearing about theistic and untraditional religious views, he could despise slavery and the class-bound English system while longing to take an elite place in it. All this was to stand him in good stead—for the next five years he was to live cheek to jowl with a real Tory traditionalist.

Fitzroy, a direct descendant of King Charles II, was as class-bound and reactionary as he could be. There was nothing the British government and lords might do that he could view as inadvisable. Darwin had to bite his tongue as Fitzroy described how happy slaves claimed to be (as their masters looked on), and appeared unfazed at the sight of slave families dragged apart for the good of the empire's commerce. It probably left a bad taste in Darwin's mouth, no doubt even with respect to the religion he was so close to taking on as his profession, for Fitzroy was as staunchly religious as he was staunchly Tory. In Fitzroy's mind, every bit of nature, every remarkable fossil or landscape could be interpreted in a scriptural context. There would be little real scientific speculation on this trip, at least not in the captain's cabin.

Darwin had his own tiny little room on board, where he kept his books and equipment. There, too, he spent day after day in seasick misery. He never adjusted to life at sea, and as much as he loved and appreciated his adventure, he knew he would never travel by ship again. He spent as much time as he could on shore, often traveling overland to meet up with the *Beagle* in some other port. By the time they turned toward home, he had walked the Amazon jungle, ridden the pampas, and trekked the Andes.

Just before the expedition left, Sedgwick had given Darwin a new book called *Principles of Geology,* by Charles Lyell. It espoused a different theory for the formation of the earth: uniformatarianism. Instead of suggesting that sudden catastrophic change was behind the creation of the earth's features, Lyell insisted the earth changed slowly. The causes were more pedestrian, just the everyday things experienced within the lifetime of every human. Everyone could see the action of waves on rocks, even if they could not measure their overall effects. Most could see what water—rivers, creeks, melting snow—could do to a landscape. And most had heard of—if they hadn't experienced directly—the effects of volcanoes and earthquakes. These were the sorts of things that had, over long periods of time, created the world's mountains and valleys. While Sedgwick believed Lyell's theory had value, he warned Darwin not to believe all of it.

Darwin looked at the world as a geologist now. When he wrote in his journal, when he collected samples, he was offering primarily geologic descriptions. He took samples of rocks; he studied the formations of the islands and mountains of South America. As he observed these startlingly new landscapes, he read Lyell's *Principles*. He found he could clearly delineate the histories of those landscapes. He formulated his own theories, such as that the formation of stepped-down plains of the pampas came from the slow draining of inland seas, with new shorelines forming sequentially. Finding fossils high in the mountains, perfectly intact as if untouched by catastrophe, made him doubt the roiling, catastrophic earth history of Sedgwick. He tried to follow Sedgwick's advice and doubt Lyell's conclusions, but more and more he saw the past as Lyell described it.

Then in Chile, as Darwin rested during a hike in the forest, he felt an earthquake. The earth, once so solid, shook "like a crust over a fluid." He raced back to the *Beagle* and they sailed north to the city of Concepción. "The whole coast was strewed over with timber and furniture as if a thousand great ships had been wrecked." The town had crumbled with the shaking, and then a twenty-foot tidal wave washed into town, carrying with it a schooner. The town was devastated and scores of people killed. Darwin was in thrall, having never imagined nature could act with such force. He walked the shoreline and saw mussels dying on the rocks, a few feet above the tide. He saw for himself how much the earth could change. There was no need, he could now see, to propose catastrophes of global scale to explain the rising of mountains.

Darwin did not spend all his time in geologic musings. He observed and collected many specimens of animals and plants as well. He started at sea, collecting plankton in drift nets. Observed under his microscope, they were astoundingly beautiful, and he mused on why there was so much beauty and variety in the middle of the ocean, where no one could admire it. In the pampas he collected two species of rheas, ostrichlike birds. These two species differed only slightly in size and in the color of their legs. They lived as neighbors, one species being predominant on the northern pampas, one in the south, with a small area of overlap. He marveled at the different forms of land animals, how different they were from African animals yet how similarly they lived.

Darwin also collected fossils, sometimes to the irritation of Fitzroy, for some of the pieces were large and heavy. Although Darwin couldn't identify them all on site, he found fossils of ancient sloths and armadillos that looked like larger versions of species still living, but singular to the continent, not represented on any other.

Besides surveying the South American coastline for the Crown, the *Beagle* was also delivering missionaries. On a previous trip, three native Tierra del Fuegans had been brought to England and civilized. These were people who dressed and acted the way Darwin was used to, people he enjoyed spending time with. Then they made landfall in Tierra del Fuego. The natives there were filthy, with matted hair and almost no clothes, even in the frozen weather. Their language was made up of what to Darwin were horrid sounds. He could scarcely see the resemblance between the pleasant missionaries and their unimproved relatives. The primitive nature of these people stunned him.

> How entire the difference between savage and civilized man is. It is greater than between a wild and a domesticated animal. . . . I believe if the world was searched, no lower grade of man could be found.

On the way home the *Beagle* stopped in the Galápagos Archipelago. The traditional view is that these dry volcanic islands were the place Darwin finally saw his way to his theory. In fact, when he got there he was mostly concerned about getting the rest of the way home. Fitzroy had had a breakdown but recovered, the trip had dragged on for three years longer than planned, and Darwin was dreaming of a landlocked parish, complete with cozy wife and rosy-cheeked children. He marveled at the tortoises and finches, the mockingbirds and marine iguanas. He collected samples with remarkable ease; the creatures would let him catch them by hand. But he didn't label them carefully; the finches were not marked with what island they'd come from, and he didn't collect any tortoises at all, even though the locals told him that each island had its own specific species. Darwin, when he got to the Galápagos, was burned out.

Still, what he saw there agitated him. On the *Beagle,* just four months from home, he mused, in nonseasick moments, in his journal:

> When I see these islands in sight of each other and possessed of but a scanty stock of animals, tenanted by these birds but slightly differing in structure and filling the same place in nature, I must suspect they are varieties . . . If there is the slightest foundation for these remarks, the zoology of the Archipelagos will be well worth examining: for such facts would undermine the stability of species.

Toward the end of the trip he'd been sending samples and dispatches to his family and to Henslow. His work had become famous and his family, on his arrival back in England, insisted he publish an account of his travels, saying his letters were better than 99.9 percent of the travelogues already in print. His father at last was pleased and proud— so much so that he'd even stopped grumbling when Charles begged for more money.

HOME

Darwin arrived home in 1836, swearing that the landscapes of England were the most beautiful in the world. He came home to his family and to the scientific establishment, which was as eager to meet him as he was eager to be received and recognized by it.

Now the real work began. Boxes of rocks, birds, animals, and plants needed to be sorted and identified, by the best experts available. Many, if not all, were eager to start, to Darwin's pleased surprise. His reputation, made by his dispatches and the boxed samples that preceded him, was on the way to being made.

To his dismay, he was forced to live in London, to be close to the museums. He was also close to his brother, Erasmus, who had given up medical practice and lived a life of intellectual adventure and leisure. He kept a sort of salon where modern young minds shared radical ideas. He also had a somewhat scandalous relationship with Harriet Martineau, a feminist and atheist writer, a woman who had met and admired Thomas Malthus.

So Charles Darwin again spent some time in a world of ideas not all associated with the natural world. As he sorted and distributed his samples and collected his notes to write his *Journal of Researches into the Natural History and Geology of the Countries Visited During the Voyage of H.M.S. Beagle Round the World: Under the Command of Capt. Fitz Roy*, he was exposed to the modern social ideas of the day.

He was also joining the scientific establishment. His collections and journal brought him the sort of attention he had hoped for from accomplished scientists. Scientists like Richard Owen, a renowned comparative anatomist, agreed to study his fossils. Ornithologist John Gould would classify the birds. Even Lyell, the great geologist whose book changed the way Darwin saw the world, wanted to meet him, not just vice versa. He began to travel in rarefied scientific circles, and was invited to join organizations. Ever friendly, ever modest, he fit in well—the humble yet ambitious young gentleman scientist. He cherished the attention and recognition. He would do nothing to disturb the view these heroes were forming of him.

As the samples he collected were classified, and as Darwin worked on his journal of the voyage of the *Beagle,* and started a geology of South America, his mind was working. Evidence was piling up for what he called the transmutation of species. The mockingbirds he collected from the Galápagos were not just varieties of South American birds as he thought, they were separate species. The dull little black birds he collected from the islands were all finches, but finches that looked like other species—like finches morphed into grosbeaks and cactus wrens. As he answered the questions of the ornithologist Gould, he had to go back to Fitzroy and his assistant to try to remember where the little birds had come from. He found out the marine iguanas did not occur in South America as he had thought; they only occurred in the Galápagos. And he recalled the comments of the sailors who told him that each island hosted it own particular tortoise. Where he had not thought so much of the islands while he was there, now the evidence he'd gathered from them spoke to him more and more clearly.

A year after his homecoming he started a notebook and named it *Zoonomia,* the name of his grandfather Erasmus Darwin's book on evolution. (He did nothing carelessly or haphazardly.) He started listing the questions driving him to think about transmutation, twenty-seven pages' worth. He postulated answers, deriving from these the sorts of experiments he would have to do and information he would have to get to resolve them. He kept it all to himself, his instinct for discretion intact.

In the meantime, others were proposing their own versions of the idea. Robert Grant was still around, although generally discounted and despised by established scientists. There was also a new book, *Vestiges of the Natural History of Creation,* written anonymously. *Vestiges* made a clear case for the transmutation of species, but was written in a popular style with dubious science. The geology in particular, Darwin noted, "strikes me as bad, and his zoology far worse." Still, Darwin didn't think it deserved the four hundred pages of outraged objections penned by his former employer Adam Sedgwick, now Woodwardian professor of science at Cambridge and president of the Geological Society. Sedgwick's comments included sentences like this: "If the book be true, the labors of sober induction are in vain; religion is a lie; human law is a mass of folly, and a base injustice; morality is moonshine; our labours for the black people of Africa were works of madmen; and man and woman are only better beasts!" One of the consequences of reviews like this was that *Vestiges* sold out, some 24,000 copies (far outstripping Lyell's *Principles* and, later, Darwin's *Origin*). And despite the "bad science," it inspired Alfred Russel Wallace and Herbert Spencer, a journalist friend of Harriet Martineau. Spencer proceeded to write his own theory of evolution, in several publications, and finally in 1862 a book called *First Principles,* which emphasized a sort of metaphysical overweening concept of evolution, which Darwin read, but never claimed to understand.

Still, the general reaction of his peers and those who were still his heroes to these books kept Darwin silent. To add to his reticence, the working classes took the ideas of evolution for their own cause. Often the scientific validity was lost in the conversion of the ideas to propaganda for the purpose of demonstrating the illegitimacy of the ruling classes' claims to power. Darwin was not of the ruling class, but he shared their distaste for the commonness of the working class. These were sufficient reasons for Darwin to keep his idea close to his chest. He did plan to announce it, but not until he had as much incontrovertible evidence as could be collected. He could hope, too, that the intellectual climate might change.

As much as Darwin lacked the moral courage to announce his idea, he proved to be too much a scientist to deny it. It took over his life. He filled notebook after notebook with thoughts developing the idea. He contacted animal and plant breeders. He even started breeding pigeons, which meant frequenting the sort of working-class clubs pigeon fanciers frequented, the sort of places he was dismayed by. He collected information and experimented, soaking seeds in salt water to see if they would survive and germinate—all designed to support any and all objections to his theory he could think of. He did this without breathing a word to anyone, at least at first. Finally he could not stand it anymore and he found the perfect confidant, Joseph Hooker. The young botanist seemed perfect, a gentleman scientist but young enough to consider new ideas. He was also thoughtful, considerate, and in awe of Darwin, wishing to emulate the life of the naturalist abroad.

So Darwin made slight suggestions of his ideas to Hooker, always apologizing for them afterwards. To his joy, Hooker did not dismiss them, nor was he appalled. He listened and offered cogent counterarguments—the perfect accomplice.

And the climate did change. Another young turk, Thomas Huxley, entered the mix. He was not an evolutionist, but he did despise the old boy network, and he spent his time lobbying for the initiation of a profession for scientists that did not require their entry into the clergy. He took as his target the venerable Richard Owen, Britain's preeminent comparative anatomist and president of the British Association for the Advancement of Science. Owen was even more concerned about his place in elite society than Darwin, and responded quickly and venomously to any threat to his own exalted reputation. His own defensiveness would prove to be his downfall. The new guard had arrived, and although they didn't know it, they were all about to become evolutionists.

3

What Darwin Said

It was 1858, twenty years after Darwin first started a note-book on evolution, and he was only just beginning to think of announcing his ideas. The times were finally right; younger ideas were starting to dominate biology and younger minds were frustrated with old, inadequate models. But the surest sign came from an old source, his mentor, Charles Lyell. Darwin had finally, in a private meeting, told him of the theory, and now Lyell was advising him to announce it quickly or be beaten to the punch. Though the honored scientist would never completely accept evolution, he was enough of a scientist to know when to step aside. He could see that biology was in the hands of new minds, and they were already toying with the idea.

One such mind in particular impressed Lyell. He read a paper in an obscure agricultural journal written by an equally obscure collector, someone Lyell called a "muddy boots biologist." This paper by Alfred Russel Wallace made a provocative kind of sense, and, he realized, sounded an awful lot like what Darwin had finally admitted he'd been thinking. For Lyell, it meant the ground was quaking under his feet, but he knew better than to try to stop it. Publish now, Lyell told Darwin, or you might be forestalled.

Darwin was ready, but pictured a huge work—at least ten volumes, filled with examples to make undeniable his new laws of biology. He was still cautious, still remembering the castigation of the previous authors of such heresies, and he wouldn't risk his own position without strong, unassailable evidence. Already the ground was prepared, partly by him, partly by the times, with the new guard eager and willing to take on new challenges. Young biologists like his friends Joseph Hooker and Thomas Huxley were looking for a cause célèbre to throw in the faces of the old, elite fogies. Darwin was preparing just such a treatise.

Wallace was to be a tentative member of Darwin's alliance. Darwin had not been impressed by the collector's first paper, the one that had impressed Lyell, finding it "nothing new" and too "creationist." (Wallace mentioned the "creation" of species, but had not meant it in the biblical sense.) He reread the paper after Lyell's warning, and saw that it was indeed foretelling a risk to himself. He wrote Wallace, recruiting him into the cadre of young biologists with new ideas. He kept him as a correspondent and paid him for contributions to his own collections, this time for specimens of domestic chickens, ducks, and pigeons from the Far East. Shipping costs, he told Wallace, were killing him, but the specimens were invaluable.

He praised Wallace for his thoughts on species; they thought alike, he told him. Still, he made sure to claim the idea for himself, making it politely clear it was his territory, and probably too complex for anyone else. He wrote:

> I can plainly see that we have thought much alike and to a certain extent have come to similar conclusions. This summer will make the 20th year (!) since I opened my first-note-book, on the question how and in what way do species and varieties differ from each other. . . . I am now preparing my work for publication, but I find the subject so very large, that . . . I do not suppose I shall go to press for two years. . . . It is really impossible to explain my view in the compass of a letter.

Wallace, struggling with fevers, rainy seasons, and finding decent transport between islands of what is now Indonesia, welcomed the correspondence. Here, after all, was a man whose works he'd read and admired, a man admired by all English science, taking time to write to him. To top it off, they had the same ideas, which had to make Wallace more assured of his own conclusions, not that Wallace wasn't already a bit cocky. He wrote to his friend Henry Walter Bates to tell him Darwin approved of his theories, and "agreed on almost every word." He added that the great man might save him from writing the second part of his hypothesis "by proving that there is no difference in nature between the origin of species and varieties, or he may give me trouble by arriving at another conclusion, but at all events his facts will be given for me to work upon."

Then Wallace wrote another paper, and sent it to Darwin first for his opinion. For Darwin, this paper meant disaster, for it appeared Wallace had hit upon the same theory. It turned out their ideas did not agree completely; the differences would be made clear later. But they did both describe the same central theme for evolution—one describing the central force driving evolution—and they did this where equally brilliant, better-trained biological minds had failed. To see why it took these two particular minds to make the same breakthrough at the same time, we should look at who Wallace was, as well.

WALLACE

In a way, Wallace and Darwin were the mirror images of one another. Darwin came from working-class origins from forebears who'd been remarkably successful and established a toehold in the upper classes. Wallace came from privileged parents who had lost their fortunes, leaving their children no choice but to make their own way. Both individuals had the sense of not fitting in, of not naturally belonging to society's upper echelons. But while Darwin claimed a vicarious membership and feared losing it, Wallace had nothing to lose. Instead of fearing to offend and disrupt, he was only too willing to upset the status quo.

Wallace's father started out with an independent income and acquired a profession, attorney, that he didn't need, and never practiced. Instead, he lived a life of leisure, frequenting intellectual and artistic gatherings. Then he married a much younger woman with a fortune of her own. Things started off well, but, as their family grew, their fortunes started to crumble. The senior Wallace proceeded to make a series of disastrous business investments. As the family lost more and more money they were forced to sell their country estate.

Wallace felt this loss acutely. He later claimed he could not remember the faces of his family but he could remember every nook and cranny of that landscape. From there the family moved to a series of ever smaller and shabbier houses. Wallace's school career was cut short for lack of funds, and he was sent to work. He became a helper in London for the building firm that employed his older brother, William.

Wallace entered a rough working-class environment. He was just fourteen and his employer and fellow workers were protective of him, so it was not unpleasant. Forever after he felt a loyalty to these people, making him a lifelong socialist. He continued to read and learn. To do this he joined a worker's club called the Hall of Science, a place where workers could gather for coffee and borrow books and magazines. Lecturers came to these clubs, some spreading revolutionary socialist doctrines and protesting the class system. After losing his chance for more traditional schooling, Wallace was excluded from the more accepted circles of intellectual thought. So unlike Darwin, he saw little risk to agreeing with radical social theorists.

He tried many jobs, including jewelry making and teaching. He learned surveying from his brother William. The work appealed to him, for it put him in the outdoors, where he could tramp through the countryside and relive the happiest days of his childhood. It also allowed him to travel throughout England. His socialist leanings were fortified by these experiences: some of the work he and William did was to redraw the boundaries of Welsh common lands for the Enclosure Act, where the claims of poor farmers to common lands were stripped away, all so the rich could take possession.

For Wallace, science was the antidote to the unfair and hidebound workings of the English theocracy. Resentment against technology brought by the Industrial Revolution had ended. While it had been seen initially as oppressive, workers could now see industry as a way out of poverty. Families like the Darwins and the Wedgwoods had gained entry into the elite classes by adopting technology; it was seen as a great equalizer. For Wallace, to study science was to embrace the worldview that brought that technology, and besides, he, like Darwin, loved the natural world. While walking the roads learning the surveying trade he collected and pressed wildflowers, becoming an amateur botanist of sorts. He met a like mind in Henry Walter Bates, another self-taught young man who had had some success as an amateur entomologist, including a paper published in *The Zoologist*. Like Wallace, Bates was training for a trade; he was an apprentice in his family's hosiery business. Both had driving interests in biology and nature, and neither could have any expectations of earning a living that way. There was no such thing as career scientist—science was the province of wealthy individuals or professionals like physicians and curates. To teach at university level would require a theology degree, which neither desired, nor could afford.

The two started making plans for their own expedition to tropical lands; they both dreamed of contributing their bit to science, along with the likes of Humboldt and Darwin. Neither had any right to hope to succeed, but two dreams work better than one in cases like these.

Bates worked on convincing his family to let him go. Wallace started surveying in earnest. With the advent of steam trains, Britain was building railroads, and for that they needed surveyors. There was no end of surveying to be done; there was so much work it ultimately meant the death of Wallace's brother William. He was out surveying in a blizzard, caught pneumonia, and died. Despite this, Wallace took as much work as he could, even recruiting another brother, John. Together they were able to save enough money to buy a new home for their mother (their father had died), and bring a sister back to England from governess work in France. Even after those expenses, the two brothers saved enough for their adventures. Alfred staked a trip to the Amazon, John a trip to California to join the Gold Rush.

Bates and Wallace planned to support their trip by collecting. Natural theology was still the rage, and the wealthy spent thousands of pounds making private collections for the glory of God and themselves. They got agents to sell their specimens and contacted the British Museum to see what that institution might want to purchase.

They prepared for the trip by reading as much natural history as they could. They read Robert Chambers's *Vestiges of the Natural History of Creation.** Bates's reaction was very establishment, very like Darwin's—he was critical, finding the science uneven. Wallace's reaction was characteristic. Instead of finding reasons not to like the book, he found the central theme exciting and argued that it made sense. He admitted that Chambers left a great many holes, but this he found inspiring. As Wallace saw it, it was up to him to plug those holes. Wallace started his own notebook, dedicated to documenting every fact he observed that would support Chambers's claims for mutable species. Wallace already knew what his role was to be: he would supply the proof that species were not, as Lyell claimed, immutable. He would be the young turk proving that species changed.

When Wallace and Bates set out for the Amazon they were just twenty-two and twenty-three years of age, but scientists in those days were adventurers. Arriving at the mouth of the Amazon, they started out collecting as a team. Adventure takes its toll: after a few months they separated; why they did remained their secret. They remained distant friends and stayed in written contact, though they were not to see each other again for years. Bates stayed there many years and became a well-known and admired entomologist.

-----

* This is the same anonymously written *Vestiges of Creation* that drew the criticism of Darwin and Sedgwick, later known to be the work of Robert Chambers, who nonetheless remained reluctant to admit to being the author.

Wallace's Amazon trip took him thousands of miles, as far upriver as one of its tributaries, the Rio Negro. He started out with local Portugese-speaking guides, but, as time went by, he ended up living with native tribes. He was nearly tireless, learning to speak Portuguese, learning the collecting trade, learning to shoot and prepare specimens. He observed and collected natural history data, finding facts to support his own theory for the origins of species. He made careful notes on the distribution of species—over and over again he saw how similar but separate species lived separated by the expanse of the river.

He spent three years in the Amazon, in the end collecting specimens with considerable commercial success; it began to look more like a profession than a young man's folly. His younger brother, Herbert, joined him to learn the trade, but found it not to his liking. Sadly, on his way home to Britain, Herbert contracted yellow fever. Wallace heard briefly that his brother was seriously ill; out of touch in the jungle, the news of Herbert's death did not reach him for months.

Thus began an ill-starred homecoming. This part of Wallace's life is so eventful, Joseph Conrad and Somerset Maugham used him as inspiration for various characters, and recently A. S. Byatt wrote a novella based on his life. On his journey out of the jungle, Wallace visited his brother's small grave, then collected some boxes of specimens meant to be shipped before, but which had stalled in customs. He loaded those boxes, thirty-four living specimens of birds and animals (including a tame toucan), many notebooks and sketches, and boxes of his own private collection, onto the ship. The toucan escaped, spun into the river, and drowned—an omen. Then just two days after setting sail, the captain of the ship came to Wallace to tell him he thought the ship was on fire. Wallace, stunned, collected only a few things—shirts and a meager sampling of his sketches and journals—placed them into a metal box, and joined the crew to the lifeboats. All the live specimens and three years' worth of journals were lost. His own private collection containing (in his words) "hundreds of new and beautiful species, which would have rendered (I fondly hoped) my cabinet, as far as regards American species, one of the finest in Europe" went down with the ship.

They were rescued some days later, but even then the ordeal wasn't over. The rescuing ship was ancient and foundering. When a storm nearly knocked the ancient ship down, the captain prepared to chop down the masts; were the boat to capsize, it would right itself and all might not be lost. The ship and all were spared. The storm abated with the ship still intact and they made it home. Wallace returned with nearly nothing. He was saved from complete financial ruin only because his agent had the forethought to insure the collections for 200 pounds.

After such an adventure, most of us would take up a safe life of surveying, but not Wallace. Bored after a few months home, he was already making plans for a new trip, this time to an even more remote region, the Malay Archipelago (what is now Indonesia). His reputation as a dependable collector was well set, and support was easier to get, although his plans changed several times. In the end, his supporters set him up with a first-class ticket and enough money to spare for him to hire a young assistant. This expedition was to last eight years, and Wallace, as he traveled from island to island, clearly saw the effects of geographic isolation on the evolution of species.

It was in Singapore, early in the expedition, that he wrote his first paper, "On the Law Which Has Regulated the Introduction of New Species," the one that set Lyell to worrying. This law, as stated by Wallace, says simply that every species has come into existence coincident both in time and space with a pre-existing closely allied species. He was more than testing the waters—to make a conclusion of common origin was implied from this law, and Lyell realized this. Wallace went on to describe the classification of species as appearing like a tree, a metaphor Darwin had used as well. Perhaps it was this repetition as well as the wording of the following phrase that allowed Darwin to dismiss the paper: "that the islands were successively peopled from each other, but that new species have been created in each on the plan of the pre-existing ones." Using the word *created* made Darwin think that Wallace was just another creationist, but Wallace was using the word for convenience, the word *evolved* not yet being in common usage.

Darwin, after rereading the paper on Lyell's advice, saw what he had missed before: the obscure collector was close to his own idea. He wrote Wallace immediately, treating the collector as a member of a special club, someone who might understand Darwin's secret ideas. He assured Wallace that he found the paper insightful and close to what he himself had been thinking, but then told him there was so much more to his own theory, he couldn't possibly tell him

53

in a letter. In the meantime, Wallace could be an ally. Darwin wrote to recruit Wallace and solicited him to collect exotic domestic chickens, ducks, and pigeons for Darwin's experiments at home.

Wallace, who had written the paper as a clarion call to announce that he had ideas, was itching for argument. Half a world away he knew he would have to wait months, but even then there was nothing. The sole response was one reviewer advising him to stick to collecting and to do a better job at it, and not waste his time with theorizing. To finally get Darwin's complimentary letter one year later was exciting and a godsend.

But while Darwin fretted and fussed over his many-volumed opus, Wallace took sick with malaria. Stuck in a hut in Ternate in the Mollucca Islands, he had a revelation. All because he remembered an essay he'd read some twelve years before.

THE IMPORTANCE OF MALTHUS

The Reverend Thomas Malthus had lived and worked in the same township Wallace was born in. He is largely held to be responsible for the labeling of economics as the "dismal science." He answered the mostly optimistic social predictions of the French Enlightenment thinkers by pointing out one not very pleasant fact: human populations grow quickly—too quickly for the resources they need to survive. If unchecked, human numbers grow exponentially, while necessary resources increase only arithmetically, since their production is based on a limited resource—land. The growth in the supply of food and other resources necessary for life would not keep up. The result would be scarcity, and humans would have to compete for those scarce resources.

Malthus's essay became well known and was used promiscuously. Social activists tended to use it to push for birth control and welfare for the poor. Laissez-faire capitalists tended to use it to argue against such measures, pointing out that scarcity made for incentive, and pumped energy and innovation into the economy. The poor would have to struggle to survive and would have to find their way to new production innovations, as Adam Smith theorized, to the betterment of the whole of society. Malthus's theory was co-opted by both sides, and bandied about until everyone was sick of it. (The theory survived this later disdain only because of its basic truth.)

It took Wallace and Darwin to apply the idea to biology. Both admitted to being under the profound influence of Malthusian thought when they came to their theories. Wallace puts the revelation most clearly:

> One day something brought to my recollection Malthus's "Principle of Population," which I had read about twelve years before. I thought of his clear exposition of "the positive checks to increase"—disease, accidents, war, and famine—which keep down the population of savage races to so much lower an average than that of more civilized peoples. It then occurred to me that these causes or their equivalents are continually acting in the case of animals also; and as animals usually breed much more rapidly than does mankind, the destruction every year from these

causes must be enormous in order to keep down the numbers of each species, since they evidently do not increase regularly from year to year, as otherwise the world would long ago have been densely crowded with those that breed most quickly. Vaguely thinking over the enormous and constant destruction which this implied, it occurred to me to ask the question, Why do some die and some live?

Darwin admitted to having a similar revelation some twenty years earlier, but he hesitated before writing it down. Wallace wrote it down as soon as his malarial fever subsided enough so that he could. He wrote it and posted it to Darwin within three days. He asked Darwin to read it, and if he thought it worthy, to forward it to Lyell.

LETTING THE SECRET OUT

Thus arrived the letter, and Darwin's moment of truth. In the end, he did exactly as Wallace asked and forwarded the letter, and with it his dilemma, to Lyell. He included Hooker in the decision as soon as he could. What should he do, he wailed to his friends, "I would far rather burn my whole book, than that he or any other man should think that I had behaved in such a paltry spirit. Do you not think that his having sent me this sketch ties my hands?"

Darwin's hands maybe, but not Lyell's or Hooker's. They could act when he couldn't, a fact that Darwin realized. It was not all expediency, for Darwin was horribly preoccupied at the time; his youngest son, Charles Waring, a sickly and retarded child, was dying of scarlet fever. Other family and household members were sick as well. Darwin himself was seldom well, spending most of his time either vomiting or prostrate with nausea. If he left to his friends the task of deciding his and Wallace's fate, he had his reasons.

Lyell and Hooker decided on a Solomon-like maneuver, but instead of dividing the baby, they would present twins. Darwin had previously written an essay (he sealed it and left it for his wife to release in case of his death) and a letter to the American biologist Asa Gray in which he briefly explained his theory. They gathered these two documents with Wallace's essay and read them at a special session of the Linnean Society, a by-then somewhat obsolete group. Several other papers were read at the same meeting, and the twin theories did not excite much reaction. Hooker believed the lack of protest was because of Lyell; association with the well-respected scientist lent the theory more legitimacy, and that quelled the sort of scathing response they would have otherwise expected. Still, the silence was deafening. Famously, the president of the society stated that the meeting had not "been marked by any of those striking discoveries which at once revolutionize, so to speak, our department of science."

By presenting both men's versions at a relatively small meeting, they bought Darwin a reprieve. He now had to quickly write his ideas for publication, which meant he could not complete the many volumes he had planned. Instead, he spent the next year feverishly preparing his *Origin of Species,* a book he insisted be called an abstract until his publisher advised him it would frighten away too many readers.

Wallace was still traveling the islands when Darwin published. When he heard his paper had been presented in parallel with Darwin's, he was gracious; some might call him naive. He wrote his mother, "I have received letters from Mr Darwin and Dr Hooker, two of the most eminent naturalists in England, which have highly gratified me. I sent Mr Darwin an essay on a subject upon which he is now writing a great work. He showed it to Dr Hooker and Sir Charles Lyell, who thought so highly of it that they had it read before the Linnean Society. This insures me the acquaintance of these eminent men on my return home." Considering the class-bound system of England at the time, this reaction was probably appropriate. The mere collector, the "muddy boots biologist," had succeeded in entering a society that, not long before, would have crumbled rather than accept such a rabble-rouser.

THE IDEA

Darwin and Wallace came to the same breakthrough idea: natural selection. What natural selection did was provide an engine for change, an external force to drive species to change, a force analogous to Newton's gravity. In a way, the two did little more than take a natural step that their predecessor, Charles Lyell, had avoided. Lyell's uniformatarianism, his conclusions that the earth changed slowly, gave them the force. Darwin and Wallace simply came to the inevitable conclusion: If the geologic environment changed, so too must the living things in it.

Natural selection means that organisms change, generation by generation, as dictated by the environments they live
in. The change comes about through the survival and reproduction of the best adapted (the most fit) and the failure to
reproduce of those not so well endowed. If the organisms are
already fit and adapted, there may be no change. But if they
are not, there will be change or, if change does not occur,
there will be extinction. These slow and imperceptible changes
will mean the organisms take on new forms—forms that ultimately mean the development of new and separate species.

Like Newton, Wallace and Darwin took an external
approach. Species did not change from within, but because
external forces (specifically, changing environments) forced
them into changing. Malthus gave them the means—they
applied his essay on human populations to all organisms,
and it set for them a problem only evolution could solve. It
was clear that all species, even the ones slowest to reproduce (such as the elephant), would overrun the earth if left
alone. If populations did not increase uncontrolled there
must be checks, and they found them: predation, starvation, and disease. These natural pressures would winnow
away numbers, leaving a few to survive. And which few? As
Wallace asked, "Why do some live and some die?" That
question focused on the why of species transmutation like
a laser. Wallace and Darwin answered by pointing out that
the ones that live must be different: faster, stronger, bigger,
smaller, or perhaps just more resourceful—in a word,
fitter—than those that die.

Several points deserve special mention in this theory.
First, there is no impetus of design in this theory. That is,

there is no overarching idea or plan; change occurs if and only if it means better survival and reproduction. There is no direction to the change; organisms do not necessarily become more complex or more intelligent unless those characteristics contribute to survival and reproduction. Second, there is no mention of how these changes occur. Genetics was an unborn science at the time of Darwin and Wallace. Third, there are some implicit requirements that Darwin recognized and knew had to be established in order to make the theory work.

He had a monumental task, which he recognized. Evolution, a word not used by Darwin or Wallace (evolution at the time meant only "to roll out, as in a scroll" and was thus not without its own implications of design and plan), was not something that could be proved the way Newton could prove gravity. Organisms ultimately would not act as predictably as did inanimate objects. Plus, there was a time element—all these changes would require periods far longer than a single human life span, thus there would be no experiment possible that would prove evolution. This was clear to Darwin. He had to invent a new approach to establish a new kind of science.

Concurrently in Europe there were physical scientists looking at analogous problems. Studies of gases and fluids were under way, and scientists were realizing they would have to predict behavior on the basis of populations of molecules. They began thinking in terms of trends. At the same time, mathematicians were working out the rules of probability and statistics. Both of these new studies would have helped Darwin, if only because they represented a new way of thinking about proof. But they were not well enough developed, and Darwin lacked the aptitude for math and physics (later on he admitted regretting the lack of discipline to apply himself to studying math as a student) to be aware of their emergence.

Even with these tools, he still wouldn't have had a complete solution. Living systems are what we now call complex systems, and as such there are few simple, linear solutions to problems concerning them. The mathematics needed to be able to predict the behavior of complex, chaotic systems is only now beginning to catch up with messy reality.

So what was Darwin's solution? Here was a theory he knew was correct, but he also knew few would accept it. The science of his time was unwilling to either believe in such an indeterminate idea or to accept a scientific proof of such vagueness. He could neither offer the clean precision of Newton nor the reassuring comfort of Genesis. He would have to bludgeon his idea through; he would have to provide "one long argument."

PROOF: THE ORIGIN OF SPECIES

Does Darwin deserve the credit he has gotten for the theory of evolution? A few historians have taken the lost month and concluded that Darwin wouldn't have come to the conclusions he had without Wallace's help. Or that Wallace supplied some answer that Darwin was struggling with.* The best refutation to these claims is Darwin's book itself. *On the Origin of Species* is Darwin's best defense. It is an impressively complete document, heavy with cogent details from organisms as diverse as lilies, barnacles, pigeons, and bees.

By organizing *Origin* the way he did, Darwin pays tribute to one of his heroes, John Herschel. Herschel was a scientific philosopher who wrote *Preliminary Discourse on the Study of Natural Philosophy,* in which he set out the requirements of scientific method, to a great extent basing his ideals on the work of Isaac Newton. Newton used a pendulum or a rock on a string as real life examples of the orbits of planets, a literary device Herschel called a *vera causa.* Darwin chose to use the same device; in his case he chose to discuss artificial selection, the breeding of domestic animals, as a real-life familiar example of natural selection.

It was a clever move. He himself, a devotee of sporting dogs and horses, understood the value of a well-run breeding program. And as a countryman, he knew that every

---

* Perhaps the critics find the latter part of Darwin's life a source for their doubts. After his five-year expedition, Darwin never again left England, hardly ever even leaving his home in Down. Nor did he work actively in the scientific circles of the day. He stayed at home, in part because he was plagued by illness, vomiting and nauseous every afternoon, sometimes too sick to get out of bed. Historians often call his disorder psychosomatic, blaming it on a nervous disposition. A few suggest he suffered from diseases he contracted in South America—he does describe, in his notebooks for the *Voyage of the Beagle,* having been bitten by a kissing bug, a carrier of Chaga's disease. Chaga's disease is a chronic condition often affecting the gastrointestinal system.

farmer in England knew the effects of breeding a good bull to a good cow. Counties bragged about their own special breeds of cattle, sheep, and dogs, each better for producing milk or wool or for depleting the stable of vermin.

Using this example established for Darwin two prerequisites for his theory. First that even within a single species individuals varied. The existence of variation was essential for his theory; differences between individuals are the raw material for evolution. Here his choice of domestic animals as an example was fortuitous, for there is much more variation in domestic species than there is in wild species. (Think of the size difference between a Clydesdale and a Shetland pony, or the shape difference between a bulldog and a borzoi.) It didn't matter that these differences were exaggerated because of the sheltered life these creatures led; the differences had to be there in the first place before human interference could produce the exaggerations. The same is true of wild populations, like the Galápagos finches. There can be no increase in the size of a beak without a few finches being born with slightly larger beaks. Then, if having a bigger beak means the birds can crack the larger seeds left over after a drought, they live longer, perhaps long enough to have offspring.

Darwin used pigeons as his experimental example. He bred them himself for a while, even though it meant he had to hobnob with the working class, an endeavor he at first abhorred. Pigeon breeding was a poor man's hobby, and Darwin had to get his birds from clubs where men drank pints, talked pigeons, and abused the upper classes. They thought of Darwin as "the governor," a nice but odd fellow, who kept asking them to breed their pigeons in bizarre ways. He would ask them to breed a perfectly good pouter with a fan tail and then took delight in offspring that looked like nothing so much as a common, wild, rock pigeon.

Breeding pigeons was far from the only experiment Darwin performed. He begged and bought specimens from everywhere, and at times found himself inundated with carcasses. He measured everything—feathers, bones, teeth, claws—to prove quantitatively that no two individuals were identical. Then he worked out the details of dispersion; he wanted to prove that animals and plants could and did drift out to islands like the Galápagos. He spent years soaking seeds in salt water for weeks and months, then planting them to see if they would germinate. Often they would, to his delight, but the test brought on a new problem, for the seeds usually sank. They wouldn't make it to land that way. So he tried mixing seeds in mud and pasting the mix to the feet of ducks, or he fed them to birds and then floated the bird carcasses again for weeks and months. Then he dissected out the seeds, planted them, and they grew! For Darwin, there was no detail too small, if it could be done in his office or on the grounds of Down, that he would not at least try to test.

Besides variation, the second prerequisite Darwin had to show was that different characteristics would be transmitted to offspring. Artificial selection again was an ideal example. It was clear from the success of animal breeders to get the type they desired that some such mechanism existed. No matter that they had no idea of how it worked; it worked. The details of the mechanism would later stymie

Darwin, and give opponents fuel for their arguments. Just what the mechanism was—packages of molecular information we call DNA—was not to be discovered by general science until well after Darwin's death. In the meantime, he postulated the existence of information packets he called "gemmules," which almost worked. All that mattered at the time he wrote *Origin* was a way to show that it happened. The development of breeds, some with shapes and behaviors highly specialized and diverse from the original species, showed that the inheritance of specific variations happened.

Darwin extended his theory to assert that all life, all species, were descended from a single common origin—he called it "descent with modification"—meaning that there was one life-generating event from which all forms of life had sprung. To defend this assertion, he used examples from morphology. The comparative anatomists (including one of Darwin's harshest critics, Richard Owen) had clearly demonstrated the existence of homologies between species. The bones of fish, birds, and mammals were developed on the same plan. The bones of fingers and wings, of hooves and claws, were perfectly analogous, if structurally different. He went on to add the development of embryos as further evidence. Scientists such as Haeckel had carefully studied development and described how avian and mammalian embryos grew and lost gills—they seemed to "pass through" the form of a fish. Out of this study came the rule "ontogeny recapitulates phylogeny." What else could explain this bizarre series of events than the notion that all species came from the same original stuff?

He went on to test his theory against what he had observed on his voyage. He had catalogued the species found on oceanic islands—the Galápagos and the Cape Verde Islands as well as others. The facts didn't fit with a creator placing perfectly adapted species in each site. Instead there were species looking very like the types of species found on the nearest continents. And there were highly mobile species, like sea birds, found widely dispersed. Then there were endemic species. Most oceanic islands lacked terrestrial mammals and nearly all lacked amphibians—species unlikely to survive long sea journeys. It wasn't that mammals couldn't live there; often these islands supported bats and human-introduced species. Indeed, introduced species frequently proved to be better adapted than natives, out-competing them and driving them to extinction. These were often animals that could travel, but commonly did not, like song birds. Then these species radiated out, filling niches where there was no competition; thus there were finches living like grosbeaks and with the beaks to match. It would take a perverse god to populate islands like this.

After establishing—to his mind, briefly—the evidence for his theory, Darwin went on to argue against what he imagined would be the strongest objections. He discussed the lack or rarity of transitional forms. Where were the beasts with half a wing? Where the three-toed horses? He answered by describing the evolution of forms as like a tree, with the existing species found on the very ends of twigs, and loads of dead branches—extinct forms—below. Unfit species would quickly be eliminated, superseded by more fit ones, and would vanish from the earth. The paucity of the same forms in the fossil record had more to do with the general paucity of the fossil record—not everything lasted long enough for us to look at it.

Darwin revised his book six times, changing it dramatically as he dealt with new objections and questions. He rewrote more than 75 percent of his sentences, and the last edition was more than half again as long. Some parts and changes either stayed or became confused as he sought but couldn't reach correct explanations. He might not have been able to explain all the mechanisms, but he remained open to challenges, testing and re-testing his ideas.

In *The Origin of Species* Darwin essentially discovered and mapped out a new science. Today we study evolution on his terms mostly; the areas he didn't anticipate precisely, like genetics, he did anticipate in principle. He left one thing out—the descent of humans—but he did that on purpose. He knew he couldn't unload too much on the world too fast. If there is only one measure of Darwin's place in history it is this book, and this book is enough.

4

Reaction

So Darwin finally took the plunge, pushed to it by the arrival of a real competitor. He finally took his chances and announced his real position to the world. Still, he tried to be as inoffensive as possible. His style remained modest, even at times apologetic. (He was criticized for calling it "my theory," and perhaps in response changed it to "the theory" in later editions.) His most glaring concession was to leave humans completely out of the mix, only once making the suggestion that the same principles could be true for humans as well.

This reticence did not reflect his frame of mind. In his private communications he clearly stated his view that humans were not separate from other organisms in undergoing this process. In his public writings he remained polite. He opted for diplomacy, not publishing facts he had collected about humans because: "I thought that I should thus only add to the prejudices against my views." At the same time, his own opinions of man and nature were firm: "Man in his arrogance thinks himself a great work worthy of the interposition of a deity, more humble and I think truer to consider him created from animals." And, "Let man visit Ourang-outang in domestication, hear expressive whine, see its intelligence when spoken [to], as if it understood every word said—see its affection to those it knows,—see its passion and rage, sulkiness and very extreme of despair; let him look at savage [Fuegan], roasting his parent, naked, artless, not improving, yet improvable, and then let him dare to boast his proud pre-eminence."

While the announcement and readings of Wallace's and Darwin's essays at the Linnean society proved anticlimactic and nearly invisible, the release of Darwin's *Origin* was not. The book sold out its first printing of three thousand books, and a second printing was quickly set up. Public and critical reaction was dramatic; opinions were, as expected, polarized.

Opinions were not divided entirely on religious grounds—there was not a division between a creationist viewpoint and a Darwinian one. Partly this was due to the scientific erosion that had already occurred in the foundations of the "special creation" theory. Darwin was not the only one who had had problems reconciling what the biological world looked like and what people could imagine a logical and sensible creator would do. So this time the idea of changing, mutable species was not dismissed out of hand. There was a scientific need for this kind of explanation, and scientists and the public were ready for it. Interestingly enough, a few religious leaders were also ready.

There was polarization on social and political grounds, as well, although it was harder to predict. Some despised his theory on the grounds that it set up life as an essentially cutthroat and warlike existence; in Darwin's world, life was nothing less than a constant, bitter struggle to survive, pitting creature against creature. Early on, some pointed out the theory had laissez-faire implications. Darwin was amused by one quote: "I have noted in a Manchester newspaper a rather good squib, showing that I have proved 'might is right' and therefore Napoleon is right and every cheating tradesman is right." On the other hand, social reformers who used Malthusian theory to push for contraception and education found its premises sensible. His brother Erasmus's companion, the social writer Harriet Martineau, was delighted with the work.

But the book never met with the scalding dismissal that Chambers's *Vestiges* or Lamarckian theory had. Whether he had planned it consciously or not, Darwin had prepped his public well. He had successfully built a reputation among his peers as a respectable and respectful scientist. The public's image of Darwin, based on his popular book about the voyage of the *Beagle,* was of a pleasant and self-effacing man, and he had done nothing in twenty years to tarnish that image. The scientific establishment knew him as respectful (he had done nothing to upset them, but had instead joined in all their prejudices) and hardworking—he had worked for eight years to complete a comprehensive and widely admired monograph of barnacles. In addition, as a landowner he had played a generous and responsible role in his community, even at times mimicking the role he had anticipated when he studied for the clergy. All of this combined gave him a public image of being a man who would not announce such dreadful ideas merely for shock value. In the minds of the public, if Darwin believed something, he must have had a very good reason.

At the same time, Darwin had carefully cultivated allies, many of them young and ambitious. To them the status quo of a theocracy in control of science was oppressive. It was their agenda to upset it, so Darwin knew they could be depended on to disturb scientific and public sensibilities with little compunction, making his points forcefully while leaving him completely out of the fray. He had assiduously befriended these people, both professionally and person-ally. He wrote to them, soliciting their aid for samples and results, and flattered them by begging their opinions. He put them up at his house, tying them to him for more than just intellectual matters.

Now, it was time for a little payback. With his theory, he was handing them an invaluable weapon—a new idea certain to meet with rejection, revulsion even, and at the same time strong enough to stand up to malicious scrutiny. It is hard to imagine that he could have implemented a better public relations campaign, had he planned it or not.

THE BULLDOG

The first line of defense (some may say offense) was Thomas Henry Huxley, nicknamed Darwin's bulldog by the London press. A few years younger than Wallace, Huxley nonethe-less conducted himself with the assured manner of a successful scientist, although both came from similarly humble origins. The son of a schoolteacher, Huxley had little formal schooling but read prodigiously on his own. With this background he managed to land a medical apprenticeship at age fifteen, then a scholarship to study at Charing Cross Hospital. He then signed on as an assistant surgeon, a position that allowed him to act as naturalist on the HMS *Rattlesnake* on its voyage to New Guinea and Australia. The frigate's main assignment, like the *Beagle*'s, was charting, with some support for biological investigations. While this allowed him to travel and study, as it had Darwin, Huxley's position had none of the privileges and freedoms allowed an upper class captain's companion. Huxley characterized his experience as follows:

> I wonder if it is possible for the mind of man to conceive anything more degradingly offensive than the condition of us 150 men, shut up in this wooden box, being watered with hot water, as we are now. . . . It's too hot to sleep, and my sole amusement consists in watching the cockroaches, which are in a state of intense excitement and happiness.

On the *Rattlesnake* he collected marine invertebrates, in particular sea squirts, and his descriptions, sent from each port of call, brought him to the attention of the scientific establishment. On return he became acquainted with such stars as Darwin, Lyell, Hooker, and Herbert Spencer. He was invited to join the Royal Society.

Unfortunately, fortune did not follow that fame. There was no career category called "scientist" at that time—even

Darwin could never have supported himself on his writings and experiments. So Huxley was forced to scrape to get by, living on a navy pension and whatever he could scrabble from classifying preserved sea squirts for the British Museum, doing popular lectures, and writing articles. The Admiralty and the Royal Society refused to pay him to write up all his researches. No university would hire him, not even with recommendations from Darwin. He lived in the least desirable part of London. Worst of all, it had been three years since his return to England and he still couldn't afford to bring his fiancée from Australia. No wonder he rebelled against the class system made up of amateur upper-class scientists: "old buffers and spider stuffers."

Early on he developed a close friendship with Herbert Spencer; both were the sons of teachers, and both had problems with women (Spencer's being a woman he couldn't shake). They lived near each other and often walked together. Spencer wrote primarily as a philosopher, but had dabbled in organic evolution. At that time Huxley could not agree that species changed, although he explained later he disagreed because he could see no evidence for Lamarckian evolution and conceive of no driving force for change. He was particularly critical of Spencer's referring to a chain of life. To even discuss life in such a way was bad science, he told his friend. Life was organized more in the form of a branching tree.

Despite being the personal embodiment of what Huxley despised—a wealthy, upper-class gentleman dabbling casually in science—Darwin quickly earned Huxley's loyalty. Darwin sent a copy of his barnacles monograph and begged him to write a review, flattering him shamelessly. Huxley returned the favor with a positive review. Although Darwin never used Huxley as he used Hooker, as a confidante and sounding board for his dangerous ideas, he probably tested the younger scientist. Certainly Huxley was primed for theory, for despite his early dismissal of Spencerian evolution, Huxley jumped on Darwin's bandwagon early.

He wrote an effusive review of *Origin* for *The Times*. Huxley not only praised the theory, he praised the way it was presented. He wrote that Darwin's book provided "a solid and broad bridge of facts" and praised him for making sure all ideas could be effectively tested with observation and experiment. Darwin had done no less than "provide us with the working hypothesis we sought. . . . My reflection, when I first made myself master of the central idea . . . was, 'How extremely stupid not to have thought of that!'"

He knew such an idea was just what was needed to shake up the old guard—perhaps even shake it down—and wasted no time in letting Darwin know he was on board. Even before the review was published he wrote Darwin saying, "and as to the curs which will bark and yelp, you must recollect that some of your friends, at any rate, are endowed with an amount of combativeness which (though you have often and justly rebuked it) may stand you in good stead. I am sharpening my claws and beak in readiness."

SCIENTIFIC REACTION

While Huxley cheered, others jeered. England's most well-read paper, the *Daily News,* was scathing, as were several other journals. Richard Owen's review in the *Edinburgh Review* shocked Darwin so much that he lost a night's sleep. Owen characterized *Origin* as an "abuse to science," calling it the sort of thing that "a neighbouring nation some seventy years since, owed its temporary degradation," this equating Darwin and Lamarck, and predicting for England a descent into the chaos of the French Revolution. He countered Darwin's criticism of creationists (asking if they really imagined animals appearing out of thin air) by asking if Darwin imagined his first progenitor flashing into being from elemental atoms.

Darwin's old teacher, the geologist Adam Sedgwick, wrote a damaging review for the *Spectator*. He also wrote Darwin directly:

> I have read your book with more pain than pleasure. Parts of it I admired greatly, parts I laughed at until my sides were sore; other parts I read with absolute sorrow, because I think them utterly false and grievously mischievous. You have *deserted*—after a start in that tram-road of all solid physical truth—the true method of induction, and started off in machinery as wild, I think, as Bishop Wilkin's locomotive that was to sail with us to the moon.

This probably didn't surprise Darwin, as Sedgwick had been an exceptionally strict don at Cambridge and, although he gave Darwin his copy of Lyell's *Principles of Geology*, he had warned him against believing it. Sedgwick took his condemnation of Darwin's theory even further, using his teaching position at Cambridge. Paradoxically, as he wrote exam questions designed to make sure his students found *Origin* as heinous as he did, the book itself was banned from the college's Trinity library. Darwin appreciated the situation, for nothing, he observed, inspired more student interest in his book than having it banned and panned.

Henslow, Darwin's champion, did not condemn the book. On the other hand, he neither counted himself a supporter. He called it a "stumble in the right direction," but also said that it wasn't really a theory but was instead an hypothesis, and that it "reminds me of the age of astronomy (before Copernicus) when much was explained by Epicycles—and for every fresh difficulty a fresh epicycle was invented."

The debate broke into the public forum with a meeting of the British Association for the Advancement of Science in 1860. The meeting was to feature a paper on "The Intellectual Development of Europe Considered with Reference to the views of Mr. Darwin." Huxley had heard that Samuel "Soapy Sam" Wilberforce, Bishop of Oxford, would be there, and though at first Huxley decided to avoid it, thinking the set-up too advantageous for his opponent, he was shamed into going. He and Hooker were worried, for it would be an ideal opportunity for Wilberforce to express the church's refutation of Darwin's idea.

They were right. At the end of the reading, Wilberforce rose to state in detail his objections to the idea, finally ending by saying that he would be distressed to be told that an ape at a zoo could be his ancestor. To drive home the dig, he turned to Huxley and "begged to know was it through his grandfather or his grandmother" that he claimed his descent from a monkey. At this Huxley is reputed to have turned to his neighbor to whisper, "The Lord hath delivered him into mine hands," then rose to say, "I would rather be the offspring of two apes than be a man and afraid to face the truth."* To make the legend even more colorful, after witnessing such an abuse of a bishop in public one Lady Brewster is reported to have fainted. Thus the lines were drawn, and the debate made news.

Darwin himself admitted that the very public nature of such controversy, lining up respectable, well-known scientists against an equally well-known man of the cloth, gave his idea currency. He wrote to Asa Gray at Harvard:

> My book has stirred up the mud with a vengeance; and it will be a blessing to me if all my friends do not get to hate me. But I look at it as certain, if I had not stirred up the mud some one else would very soon; so that the sooner the battle is fought the sooner it will be settled; not that the subject will be settled in our lives' time. It will be an immense gain, if the question becomes a fairly open one; so that each man may try his new facts on it pro and contra.

Huxley was important in other ways. He was a dynamic speaker, but even more importantly he actively tried to educate the working class. He started a series of public lectures

---

* All record of this debate is hearsay. Huxley in his own report in a letter to Darwin gave himself a far less pithy and effective response: "If, then, said I, the question is put to me 'would I rather have a miserable ape for a grandfather, or a man highly endowed by nature and possessed of great means and influence, and yet who employs these facilities and that influence for the mere purpose of introducing ridicule into a grave scientific discussion'—then I unhesitatingly affirm my preference for the ape."

that became extremely popular. (Owen, seeing their success, tried the same tactic. Unfortunately the aristocratic demeanor he had cultivated worked against him.) Huxley eagerly employed the political implications of the theory of evolution, using it to deny any sort of god-given privilege to the upper classes. Taking away from the theocracy their noblesse oblige, Huxley set up a world where success and riches could be rightfully earned by resourceful individuals. Huxley was there to let the working class know they did have a place in the modern world and to encourage them to take it. And Huxley was busy building a new profession— that of professional scientist. He groomed young biologists, encouraging them but setting strict standards of behavior as well. (Later on, if any of those same young biologists rebelled against his standards, he acted very like the sort of hide-bound traditionalist that he had initially abhorred.)

Huxley scored a major victory for Darwin against Owen in 1861. The arrival on British soil of the first examples of the gorilla did much to support Darwin's theory. The clear similarity of the ape's anatomy to human anatomy was persuasive. But Owen concocted a solution to this threat, claiming there was a structure in the human brain, the hippocampus major, that was not present in ape brains. This hippocampus was to be his swan song, for Huxley was able to publicly prove Owen wrong. This became a public victory when the following poem was published in the *Punch:*

> Next Huxley replies,
> That Owen he lies,
> And garbles his Latin quotation;

That his facts are not new,
His mistakes not a few,
Detrimental to his reputation.

"To twice slay the slain,
By dint of the Brain,
(Thus Huxley concludes his review)
Is but labor in vain,
Unproductive of gain,
And so I shall bid you 'Adieu'!"

Flush with victory, Huxley, impatient with Darwin's reticence, published his own book, *Evidence as to Man's Place in Nature,* extending the mechanism of evolution to include humankind. Darwin was working on his next book, *The Descent of Man.*

Darwin's theory was helped by another discovery. In the same year, a fossil was discovered in Bavaria of an archaeopteryx, an ancient bird. Darwin postulated birds evolved from reptiles, and this creature had the feathers of a bird but the tail and vertebrae of a reptile. It helped to support Darwin's assertion that intermediate forms existed, but just hadn't been discovered yet.

On the other side of the debate, Darwin's theory took a tremendous hit from Professor William Thomson of Oxford College, later Lord Kelvin. When Thomson announced his estimates of the age of the earth, Darwin was sharply disappointed. Thomson calculated, using characteristics of heat dissipation, that the earth could be no older than 100 million years. Darwin had already announced that evolution as he figured it required at least 300 million years to have occurred. (Later it was found that recently discovered radio-activity would delay heat dissipation, pushing estimates closer to Darwin's.)

OPPOSITION

Darwin wrote *The Origin of Species* to propose a theory more plausible than special creation. His goal was to demonstrate that the observed world was better explained by a mechanistic selection by nature than by convoluted justifications for the apparently capricious designs of a deity generally represented as sensible.* Special creation advocates attacked *Origin* strenuously, yet seldom offered much evidence in support of their own theories. They were in the position of having a newborn idea, as yet incompletely developed, as a target, and they eagerly pressed their advantage.

However, many religious leaders were willing to accept and even support Darwin's theory. They were prepared, in part, because of advances in geology; strong evidence had already forced them to accept a new physical history

---

* A book written about the same time by Philip Gosse, *Omphalos,* attempted to explain such things as fossils and Adam's navel as being the result of God designing the world to have the *appearance* of great age. He wrote that it was God's intention to design the universe with a circular theme—that beginnings and endings were supplied to make the world look as if it were constantly renewing itself. It was a convoluted and difficult book, described by the author's son as a regrettable folly.

for the earth. They were willing to extend the theistic nature of their beliefs to include a mechanistic basis for life as well. It was possible, they believed, to see creation as an extended event, and the description in Genesis as more or less symbolic.

Darwin made this support harder when he wrote *The Descent of Man*. For theologists, making humans just another evolved animal was hard to take. There were compromises. Wallace, in fact, concurred with Darwin as to the evolution of the human form, but tweaked the theory a little bit. At some point (he was not specific) the creator must have come along and breathed a soul into humans. Here was a point where Darwin and Wallace disagreed, if good-naturedly. And Wallace's writings made it easier for clergymen to continue to support Darwin's theory.

Darwin was disappointed in Wallace's conclusions, but not distressed. But the defection of another disciple, St. George Jackson Mivart, was, in his estimation, particularly painful. Mivart, a Catholic biologist and a fellow of the Royal Society and the Linnean Society, was excited and enthusiastic about evolution after first reading *Origin*. Huxley and Darwin were particularly happy to have an openly religious, as well as scientifically literate, supporter, and Mivart and Huxley became friends. Unfortunately Mivart had a crisis of confidence upon reading Darwin's *Descent of Man*. To see that Darwin and Huxley wanted to include mankind proved too much for him. He did not disagree privately, but instead wrote many articles, letters, and a book, *On the Genesis of Species,* to outline what he saw as unscientific about natural selection. In particular, there was no adequate explanation for human intellect and morality. "Science," he wrote in a letter to *Nature,* "convinces me that a monkey and a mushroom differ less from each other than do a monkey and a man."

Such criticism set a new task for Darwin. After *Origins* he had to turn from what he started with, an attack on special creation as an adequate explanation for life forms, to defending his very vulnerable theory. While he had Huxley and his young guard to lead the public fight, he was left to continue the battle in the background. He continued to do what he had done for the past twenty years, putter about in his study and garden, running more private experiments, cataloguing more evidence—all in the service of his idea. He wrote prodigiously, if slowly, producing books on orchids, insectivorous plants, plant movement, and finally earthworms, as well as *The Descent of Man* and *The Expression of Emotions in Man and Animals.* He revised *Origin* five times—the sixth edition had three-fourths of the sentences rewritten and was one-third longer.

He could no longer simply point out how much better an explanation natural selection was; he had to find ways to explain the mechanisms of his idea, such as inheritance and variation. This was a far greater challenge. Darwin's greatest talents were for observation and organization—he once described his gifts as being those of a good secretary. This was a sharp assessment, for much of what Darwin did was collect and organize information. Still, he sold himself short; he did fail to come up with the answers he needed and he did muddy the field a bit with his attempts, but he was close.

He postulated the existence of "gemmules," small particles of information that could be generated in an altered limb or organ and passed to the germ cells. (Notice that this is an explanation of Lamarckian inheritance.) He guessed these particles would be molecules or groups of molecules. He proposed that the transfer would be via the blood. His cousin Francis Galton aided him in this area of his research. He attempted to test the viability of this pathway by giving blood transfusions to pregnant rabbits. Despite the survival of most of the rabbits, he was never able to establish any tendency of offspring to resemble the blood donor.

Ruling out blood as a mode of inheritance wasn't entirely defeating for Darwin. Most models of inheritance at the time suggested the mechanism borrowed on the theme of a mixing of parental blood. The intuitive result was a model of blending inheritance: offspring receive and exhibit a mix of both of their parents' traits. With this model, children of a tall father and a short mother grow to some intermediate height. This posed an insurmountable problem: if all offspring tended toward medium, where would the variation necessary to create entirely new types (and thus species) come from? Successive generations would result in more and more similar individuals, making the development of new species impossible. Finding that inheritance didn't come from blood didn't solve the problem of blending, but it made other modes intuitively more possible.

As he grew older, Darwin kept more and more to quieter, easier experiments. He became interested in plant movement and measured the progress of plant tendrils day by day. He studied orchids, using them to elucidate and demonstrate evolutionary traits of sexual reproduction. He solicited information from missionaries and anthropologists all over the world for his book on the expression of emotions in man and animals. Even his final book, on earthworms, was a careful and brilliant treatise on the power of natural selection to even remodel the landscape itself. Not only was he the first evolutionary biologist, in many ways he was also the first ecologist.

Still, he could never eliminate the most bothersome problems. How were characteristics passed on? How did variation come about? In the words of Edward Drinker Cope, the problem became "origin of the fittest." Did new species come about suddenly or gradually? Darwin continued to be convinced evolution must be gradual, but never proved it to his or anyone's satisfaction. It has remained one of the central disagreements to this day.

DARWIN'S LIFE

Sometimes progress can be accepted only when the innovators seem least likely. Society can accept change only if it seems as onerous to the innovator as it does to the whole. So maybe only Darwin, a man who so publicly cherished his reputation as gentleman scientist, could convince establishment science that there was no way to avoid accepting evolution as fact. Here was a homebody who relished a quiet life while his ideas set the entire Western world—and much of the rest of it—on its head.

After circumnavigating the globe and returning to England, Darwin never left his native soil again. Indeed, after marrying his cousin, Emma Wedgwood, and settling in Down, he had to be strenuously persuaded even to leave home.

In part this was due to his poor health. Darwin did not understand his own illness, and secretly suspected it to be hereditary. He indulged in quack cures, attending the clinics and hospices of the most celebrated (faddish) doctors until they either failed to improve his condition or, in one unfortunate case, were forced to quit because of scandal. In the meantime, he worried constantly about his children. He feared they would inherit his weakness and suffer even more, since he had continued the upper-class tradition of inbreeding and had married his cousin. When his eldest, adored daughter Annie died, he suffered tremendously for the loss and more than once blamed it on himself.

Annie's death is blamed in part for his final abandonment of any belief in a creator. He had probably mostly abandoned at least the tenets of organized religion much earlier. He had been raised by an atheist (although his father waited until Charles was full-grown to reveal this to him), and his mother's family had been Unitarian, a fairly rebellious sect. His intent to join the clergy had more to do with his desire to pursue a career that would allow him time to dabble in natural science. He did try to maintain some semblance of faith for his wife, Emma, who was quite devout and worried constantly for her husband's faith.

Perhaps it was because of his sensitivity to his beloved wife's faith that kept him from writing on anything but science. Darwin was unusual in that, unlike Wallace and Huxley and most other writers and thinkers of the day, he refrained from religious, political, or philosophical commentary. Maybe he was just being sensible. A letter he (allegedly) wrote to Karl Marx states:

> Moreover though I am a strong advocate for free thought on all subjects, yet it appears to me (whether rightly or wrongly), that direct arguments against Christianity and Theism produce hardly any effect on the public; and freedom of thought is best promoted by the gradual illumination of men's minds, which follows from the advance of science. It has, therefore, been always my object to avoid writing on religion, and I have confined myself to science.

PASSING THE TORCH

Evolution did what all good ideas should do: it started a flurry of investigation. Various scientists in various disciplines began to test its merits within their own fields, using the theory to direct their work.

Darwin's cousin Francis Galton was, for the most part, an energetic supporter. Galton was a talented mathematician, and Darwin had solicited his help early on, having him check any calculations included in the first edition of *Origin*. (Darwin later wrote to thank Galton for finding some potentially humiliating errors.) Galton responded by praising the theory and devoting his own time and effort to its support. Besides his experiments with rabbits, he began applying mathematics to heredity. (He was passionately devoted to statistics, even gathering data to test the effectiveness of prayer. He compared the mortality rates of shipwrecks from ships carrying missionaries to those of ships not carrying missionaries.) He measured the frequencies of human traits, particularly intelligence, using questionnaires and newspaper reports. (He is often called the father of intelligence testing.) He was able to build a model of inheritance that predicted the amount of influence of parents, grandparents, and even race on the characteristics of offspring. As might be guessed, he was an early advocate of eugenics.

Galton was the first of a tradition of biometricians—biologists who used math and statistics to test and predict biological theory. This tactic was to prove divisive later, since its advocates were reviled by those biologists who persisted in remaining true to the naturalist tradition, whom they in turn reviled.

Technology drove research even in those days. Microscopes improved continually, allowing the examination of smaller and smaller organisms and parts of cells. The introduction of stains marked a great advance, mostly borrowed from textile production. By staining bodies in the center or nucleus of cells using a dye called chromatin, Walther Flemming was first to identify chromosomes. He discovered that all cells had chromosomes, and that all cells of one species of organisms had the same numbers of chromosomes. He witnessed and described these bodies clumping together then splitting apart, which later led to the cells splitting. He called this process mitosis. He speculated that chromosomes played a role in inheritance. He suggested, wrongly, that unequal splitting of chromosomes might be the mechanism behind the production of new species.

Flemming's work helped to place the means of inheritance within the cell. Another scientist, Leopold Weismann, made this pronouncement, and placed upon the cell the responsibility for evolution itself. In a powerful essay he called attention to the "continuity of germ plasm," the notion that cells are essentially immortal since they pass on their contents to daughter cells indefinitely. He extended the theory out in the other direction, noting the "cellular origin of cells," thus limiting the origin of life to other life forms. Cells became the basis and the model for all life; multicellular organisms were based on the mechanisms— and thus the inheritance mechanisms—of the single cell. Perhaps this insistence on the cell as the unit of life made Weismann adamantly opposed to Lamarckian inheritance. To prove it didn't exist, he cut off the tails of 1,600 mice for twenty-two generations with the result that not one mouse was born without a tail.

Weismann's method for this experiment was criticized because he had formulated his theory first, instead of doing an experiment and then formulating a theory from the results. He responded, saying: "To go on investigating without the guidance of theories is like attempting to walk in a thick mist without a track and without a compass. We should get somewhere under these circumstances, but chance alone would determine whether we should reach a stony desert of unintelligible facts or a system of roads leading in some useful direction; and in most cases chance would decide against us."

Darwin believed that discoveries in developmental biology offered strong evidence for evolution, although recent developments had been used against his theory. Embryonic development, once witnessed, presented a perplexing scenario. Every embryo examined, for example, whether fish, fowl, or mammal, grew gill slits at one stage in its development. In fact, at certain stages all embryos looked alike, and only developed their own species-specialized characteristics later. Developmental biologists proposed a theory from these observations, called transformationism. The transformations were the stages each embryo went through, and were believed to reflect all of life—the progressive development of more and more advanced forms. This theory was a flashback to the "chain of being" of Aristotle's day. Each organism was believed to pass through the adult form of every organism lower on the chain of life than itself. In that way a human would pass through a stage of being an ape prior to advancing to human status.

The spoiler was Karl Ernst von Baer. He pointed out that no embryo took on the specialized organs of the adult of any other species: at no time did a human embryo develop the characteristics of an adult fish. He ridiculed the transformationists in the following paragraph:

Let us imagine that birds had studied their own development and that it was they who investigated the structure of the adult mammal and of man. Wouldn't their physiological text book teach the following: "Those four and

90

two legged animals bear many resemblances to our own embryos, for their cranial bones are separated, and they have no beak, just as we do in the first five or six days of our incubation; their extremities are all very much alike, as ours are for about the same period; there is not a true feather on their body, rather only thin feather-shafts, so that we, as fledglings in the nest, are more advanced than they will ever be . . . And these mammals that cannot find their own food for such a long time after birth, that can never rise freely from the earth, want to consider themselves more highly organized than we?

Von Baer never accepted evolution, but he agreed with Darwin on the branching scheme organizing all life forms— that all were equally adapted, equally organized to their own degree. Von Baer's law of development stated that "the embryonic vertebrate, at every stage, is an undeveloped and imperfect vertebrate, it can represent no adult animal whatever." He went on to divide vertebrates into four groups based on their early embryonic forms. It was this homology among forms that Darwin used to support his theory of common origins. As the embryos developed they veered off from common pathways. In his words: "The less difference of foetus—that has obvious meaning on this view: otherwise how strange that a horse, a man, and a bat should at one time have arteries running in a manner that is only intelligible in a fish! The natural system being on theory geneological, we can see at once why foetus, retaining traces of the ancestral form is of highest value in classification." This was clearly evidence for his theory of common origins: "Community of embryonic structure reveals community of descent."

The concept of a complexity that accumulated in a linear fashion died hard (and may not even be yet dead). A successor of von Baer, Ernst Haeckel, was to breathe new life into it, and was ultimately responsible for the worst abuses of Darwin's theory. He was a great coiner of phrases, in particular his biogenetic law "ontogeny recapitulates phylogeny" (development of the embryo follows development of species). From this he concluded that phylogeny drives ontogeny. He was again imposing a hierarchy of life forms, and stating that new species were built on the forms of old—they simply added complexity and improvements. Haeckel was a strong supporter of Darwin (although the reverse was not true). Haeckel would not have historical significance on the basis of his science, since his ideas were essentially disproved fifty years before by von Baer. But Haeckel went on to suggest his theory had social implications, eagerly extrapolating it to the hierarchical development of human races and cultures. His beliefs allowed him to assert: "The lower races . . . are physiologically near to the mammals—apes and dogs—than to the civilized European. We must, therefore, assign a totally different value to their lives." His work was read and adopted by the likes of Adolf Hitler.

SPONTANEOUS GENERATION

Darwin wrote and thought on the origin of species, attempting to explain how the diverse numbers of species existing came about. He was careful to limit himself to this problem, eschewing religion and philosophy. He also refused to speculate on the origin of life itself. Once, when asked, he wrote, "It is mere rubbish thinking of the origin of life; one might as well think of the origin of matter."

This is an odd statement from someone who clearly theorized that all life sprang from one source. It is equally odd considering that his theory implied that no deity was necessary for the origin of species, so moreover none was necessary for the initial event either. It was not that the idea wasn't bandied about at the time.

Darwin lived before the germ theory of disease was proved and he lived during a time when spontaneous generation was still generally accepted. While no educated person believed you could conjure up mice in bottles filled with wheat and plugged with a soiled shirt, or flies from rotting meat, microscopic organisms were considered fair candidates for materialistic, molecular origins. One such theorist was Henry Bastian, once a member of Darwin's young guard; accepted at first by Darwin and Huxley, he became a tremendous thorn in their sides.

The problem was, if new materialist creations were popping into being all the time, evolution had to be starting over all the time. This didn't fit with Darwin's model and his belief that all species were evolved to equal states of fitness, that you couldn't rate them on a scale of *more* or *less* evolved. Darwin guessed that this sort of abiogenesis had occurred, but millions of years ago when conditions were right; and conditions were no longer right.

There was another problem. Another supporter, John Tyndall, a physicist who was also a great popularizer of science and a great admirer of Darwin, hated Bastian. In particular, he hated that Bastian didn't accept another physical scientist, Louis Pasteur, a chemist, and his idea that microscopic life came from germs (seeds) in the air. Nor did he believe that germs caused disease.

The debates went on for decades, and the protagonists wrote scathing editorials and performed competing experiments. Pasteur started by boiling a broth in a flask with a swan neck, so dust would settle out before it got to the broth. Bastian claimed Pasteur's boiling destroyed the necessary molecules for life and got his organisms by adding crumbled cheese (which contained heat-resistant spores of clostridial or tetanus bacteria) and getting turbid solutions. Tyndall finally won the day, boiling Bastian's cheese and broth solution multiple times to end up with an indefinitely sterile solution. He proved to the satisfaction of most that life came from living matter, and not inanimate molecules. No doubt it had once; it could no longer.

SALTATION

When Darwin wrote *Origin* he included the phrase *"nature non facit saltum"*—nature does nothing by jumps. This statement seemed unnecessary and dangerous to Huxley, who advised Darwin that he had "loaded yourself with an unnecessary difficulty." This was an insightful judgment, since this insistence on gradual change proved to be a persistent point of contention. People are still finding this rule hard to swallow.

Darwin, no doubt, had uniformitarianism in mind when he wrote it. He knew of the existence of "sports," the occasional birth of mutants or monsters, but did not believe them to be important in the production of new species. On the other hand, he could not offer a solid theory for where more gradual change could come from.

As a result, many of his followers rejected this rule. His cousin Galton, with his mathematical analyses, became convinced that change had to come from "jumps," or mutations. Huxley also wrote opinions supporting sudden change. Plant physiologist Hugo de Vries cultivated 50,000 evening primrose plants, producing 800 that he concluded, from their novel characteristics, must be new species. He considered them mutants and wrote *The Mutation Theory*, in which he postulated three classes of mutations: progressive, retrogressive, and degressive. This set up another line of contention, so that Darwinism became split between the "saltationists" and gradualists, as well as the mathematically inclined and the more traditional naturalists.

The theory was beginning to have a life of its own, apart from Darwin. It was being co-opted by younger scientists who were each pulling in different directions. No scientists demonstrate this more than William Bateson and Walter Frank Raphael Weldon. They met while undergraduates at Cambridge; Bateson later admitted that Weldon had introduced him to his life's work. While the two shared a passion for biology and both were as passionately interested in elucidating the origin of variation, they had profoundly different talents and ideas. Bateson admitted to having little to no understanding of mathematics, and Weldon was an eager follower of Galton and one of the first biometricians. Weldon was to join the cadre who insisted that biology would not be a science until it subjected itself to statistical tests; Bateson disagreed. But their real disagreement was about the pace of evolution, and this feud really caught fire after Bateson discovered the writings of an obscure Austrian abbot, Gregor Johann Mendel.

At the end of the century, Darwin's theory maintained a curious status. Many of its strongest opponents were seemingly willing to accept it. In the words of Reverend Charles Gore (soon to be Bishop): "In part it had been through the theologians abandoning false claims, and learning, if somewhat unwillingly, that they have no 'Bible revelation' in matters of science; in part it has been through its becoming continually more apparent that the limits of scientific 'explanation' of nature are soon reached; that the ultimate causes, forces, conditions of nature are as unexplained as ever, or rather postulate as ever for their explanation a Divine mind."

On the side of science there was no complacency, only rivalry. Biologists were drawing their lines in the sand— jumps in evolution versus gradual change, a statistical approach versus a natural one. These factions were to last, and were responsible for nearly a half-century of delay.

5

Pea Plants, Flies, and the Modern Synthesis

This was the state of biology of May 8, 1900, in the words of William Bateson:

> We want to know the whole truth of the matter; we want to know the physical basis, the inward and essential nature, "the causes" as they are sometimes called, of heredity. We want also to know the laws which the outward and visible phenomena obey.
>
> Let us recognize from the outset that as to the essential nature of these phenomena we still know absolutely nothing. We have no glimmering of an idea as to what constitutes the essential process by which the likeness of the parent is transmitted to the offspring. We can study the processes of fertilization and development in the finest detail which the microscope manifests to us, and we may fairly say that we have now a thorough grasp of

the visible phenomena; but of the nature of the physical basis of heredity we have no conception at all. No one has yet any suggestion, working hypothesis, or mental picture that has thus far helped in the slightest degree to penetrate beyond what we see. The process is as utterly mysterious to us as a flash of lightening is to a savage. We do not know what is the essential agent in the transmission of parental characters, not even whether it is a material agent or not. Not only is our ignorance complete, but no one has the remotest idea how to set to work on that part of the problem.

Not a pretty picture, but a timely summation, as things were about to change—not because of anything new but because a key piece of the biological puzzle that had been swept under the rug for thirty-four years was about to be rediscovered. Three Europeans were about to reread a single paper and see that it did indeed offer a solution to inheritance and to its vexing role in evolution. Pea plants and a chubby monk from Austria were about to make their entrance.

GREGOR MENDEL

In the few short years since the publication of Darwin's *Origin,* no doubt due to the strenuous efforts of Huxley, science was no longer trapped within the province of religion. Having religious ties had even become a handicap for the twentieth-century scientist. The fact that Gregor Mendel was a monk probably contributed to his work being ignored for nearly half a century. This was unfair, as well as foolish, because Mendel was a monk for the same reasons that Darwin almost became a country parson—it was a profession that would allow him to study and teach science.

Mendel's reasons had even more desperate roots than had Darwin's. His parents were poor peasants, and born in 1822 in Austria, he was lucky to receive any education at all. He took his vows as a Franciscan monk when he was twenty-five and was sent to the University of Vienna to study mathematics and physics. These were his first loves, and he taught physics primarily for the rest of his life. Having these disciplines for a foundation fortified his subsequent forays into biology; he applied the same rigorous analytical principles to his work with plants.

Why a math and physics major would end up changing science history with pea plants we will never know, since nearly all of Mendel's personal papers were destroyed. We do know that while he was at the University of Vienna, one of his instructors was a botanist named Franz Unger. Unger was remarkable for having developed his own theory of the evolution of plants. He published a book in 1852 titled *An Attempt of the History of the Plant World*, and in it was a chapter titled "The Origin of Plants; their multiplication and the origin of different types." In this chapter he theorized that all varieties of plants came about through changes within the plants themselves. He even postulated that all plants must have come from some single early plant, probably an alga. He gave a name to this theoretical plant, calling it an *urplanze*.

Once settled at the Abbey of St. Thomas at Brunn, Bohemia (now Brno in the Czech Republic) in 1852, Mendel devoted his life to science. He kept meticulous records of the daily temperatures, humidity, and even the level of the water in the abbey well (as a record of atmospheric pressure). He watched and recorded the activity of sunspots. He kept bees and mice, and probably tried breeding experiments with them. He founded and nurtured a local natural history society. And he started growing pea plants in the monastery garden.

There are those who claim Mendel set out to prove Darwin's theory of natural selection wrong. If he was focusing on Darwin, it was probably not selection that he concentrated on, but on another, incorrect assumption of Darwin's. In Darwin's attempts to find an explanation for variation among organisms, he postulated that the female gamete (the egg) could be fertilized by more than one male gamete, or sperm. This was something that Mendel may have been attempting to disprove.

Whatever his goal, Mendel's accomplishment set a new standard for biological experimentation. Others had grown and crossed plants, even the very same pea plants. He didn't just grow plants, cross them, and record what happened; he carefully set up a hypothesis, ran the experiment, and then analyzed the results. In the end, his meticulous methodology allowed him to conclude and report more on inheritance than any of his predecessors.

Importantly, he picked only plants with traits that didn't change. It took him two years to pinpoint these plants; he started with thirty-four plants and of these selected twenty-two varieties. These were the plants that bred true—when he self-fertilized them they produced plants that looked exactly like the parent. Then he painstakingly catalogued the traits. There were seven in all:

1. seed color: yellow or green
2. seed coat color: gray or white
3. seed shape: smooth or angular
4. pod color: green or yellow
5. pod shape: smooth or constricted between seeds
6. placement of flowers: along the stem or at the ends
7. length of stem: short ($3/4$ to $1^1/2$ feet) or long (6 to 7 feet)

Then, when he had determined his plants were pure and would always exhibit the same traits, he started mixing them up, carefully keeping track of each plant and each trait. He did this for eight planting seasons and about 30,000 plants. (Just imagine those lab notebooks.)

To be sure his crosses were pure, he had to maintain complete control over the plants. To do this, he had to open the flowers before they opened on their own, snip off the anthers—the parts that produce pollen—so it couldn't fertilize itself. Then he used a camel-hair brush to apply pollen from the plant of his choice. Then he wrapped each flower in paper or calico (cotton cloth) so nothing else—pollen-polluting bees, for example—could get to the flower. Then he waited for seeds; collected them carefully, making sure he knew which parents they came from. Then he planted them and waited to see what grew.

The results were deceptively simple; anyone who has taken an introductory biology class has seen the diagrams. Other biologists had produced and reported exactly the same results, but had come to no conclusions about them. The other biologists were not stupid, and many were better botanists. (As a botanist, Mendel made a great physicist. He called the plants he produced hybrids when they were just crosses between varieties and not hybrids at all.) The glory of Mendelian genetics (remembering that Mendel himself had no clue about genes—he called whatever was passed from parent plant to seed an "element") comes from his experimental approach. Mendel didn't just describe his results, he subjected them to statistical analysis.

The first generations of pea plants were uniform—that is, each individual cross produced plants that all looked alike. The crosses looked different from the parents and each cross looked different from other crosses, but in each "family" the offspring looked the same. Modern biologists would describe it as producing a uniform F-1 generation. In that F-1 generation each trait is preserved intact. The new plants might have a new combination of traits, mixing the various traits of their parents, but each trait would be as pure and unequivocally expressed as it was in the parent. There were no partly wrinkled peas or yellow and green spotted ones. There were no medium short or medium long stems. In simple terms, Mendel did not see any averaging of traits, no blending of inheritance.

But of course Mendel didn't stop there. He bred a second generation, an F-2 generation, this time self-fertilizing the F-1s. He hit paydirt. This generation didn't act anything like the first. Now he got variety, and he saw plants with traits that the first generation had left behind. Where in the F-1 generation there were no green seeds, no smooth pods, no short stems, in F-2 he got plants with these traits. And again, the traits were as pure and true in this generation as they had been in their grandparents. Again, there was no evidence for averaging or dilution.

These recovered traits did not show up in all the same plants. There were plants that had all F-1 traits but one, or some, or most; the long-lost traits didn't organize themselves in any conveniently located fashion. So the mathematical analysis wasn't clear and obvious; not on the first go-round, anyway.

When Mendel did sort through these results, listing and counting the seven traits, he found, over and over again, the same mathematical relationship. He found that the skipped traits showed up one quarter of the time—across all the plants, all the traits. In the F-2 generation one quarter of the plants would exhibit a trait that had not showed up the generation before. The rest of the F-2 plants would still exhibit the traits of the F-1 generation. Where in the first generation there were no green seeds and no smooth pods, now one quarter of the plants would have green seeds, and one quarter (not necessarily the same quarter) would have wrinkled pods. The more plants he produced, the closer and closer the number of these reverting plants got to being exactly one quarter of the total. From this he derived a famous ratio: 3:1. Mendel did this over and over again, producing and sorting through hundreds of thousands of plants. Every time he got the same proportions of plants; in fact, the more plants he produced, the closer his numbers got to a perfect ratio.*

If you are a physicist and you get consistency like this, you call it a law. Mendel, true to his background, did just that.

He called his law the "law of combination of different characters"; it's now known as the law of independent assortment. Mendel built this law from a simplification of what he saw happening, and didn't attempt to decipher the actual mechanism, wisely enough. Instead he just gave the materials of inheritance a name: elements. The elements that showed up in the first generation he called dominant. The ones that showed up one quarter of the time in the second generation he called recessive. This meant that anytime the element for yellow seed color combined with the element for green seed color, the seed would end up yellow. The element for green, then, would be recessive, present in the plant but not visibly expressed.

This one simple conclusion explained much about inheritance that had so far confounded evolutionary theory. Each offspring could get only one element from each parent; that meant each parent made an equal contribution to the

---

* There were biologists, in particular Ronald A. Fisher, who claimed Mendel's results were too perfect, which meant they suspected he may have fudged the data.

offspring.* The elements had no effect on each other—they remained as purely expressed when they showed up generations later as in the first wave. It ruled out in a stroke the rules Frances Galton had worked out, that offspring reflect contributions in ever diminishing quantities from parents, grandparents, great grandparents, and so on.

The law itself implied something else. Whether a trait showed up or not was determined by chance. Mendel could tell from the math that the elements must have sorted themselves out randomly. The traits fell where they might, as if someone were throwing seven sets of coins. Look at it on a trait by trait basis. If you toss one set of coins with two faces, *A* and *a*, you will get these combinations: *AA*, *Aa*, *aA*, and *aa*. If *A* is dominant, it will be the only face that shows in the mixed segments, so *AA*s, *Aa*s, and *aA*s will all look like *A*s. Statistically, each combination will come up one quarter of the time. What you would see is three *A*s for every *a*, a ratio of 3:1. In the case of the pea plants, *A* might mean plants with yellow seeds, or short stems, or constricted pods, while *a* meant green seeds, or long stems, or smooth pods.

All this would have been a great boon to Darwin had he known about it. For these results meant that inheritance is dished out in small, indivisible packages, not stirred up and

---

* This was a new and very important concept. Even Darwin had been convinced by discussions with breeders that more than one sperm from a male could fertilize an egg, which implied more male contribution than female to offspring, not a big surprise in paternalistic England.

reconstituted in a mixing bowl. So if inheritance was *particulate,* organisms were not doomed to approach some sort of average—variation could show up as if out of nowhere. Wacky traits could hide in a species for generations, just waiting for some chance combination to come along that would allow them to suddenly express themselves, perhaps in advantageous (or especially disadvantageous) ways. Maybe birds would develop a revulsion to green peas—being used to only yellow peas—leaving green pea plants to survive and reproduce better.

When Mendel's work was rediscovered, initially by three Europeans in 1900, it should have provided great reassurance to Darwinians. Particulate inheritance meant significant change could happen within the normal mechanisms of inheritance. Mutation wasn't needed for variation. Instead, this discovery added fuel to the fire of the argument over gradual versus saltatory (discontinuous) change. Some insisted it meant Darwin was wrong about gradual change—evolution occurring due to the slow accumulation of small bits of variation—claiming Mendel's peas proved change was not gradual at all, since each generation differed in a qualitative fashion. It took decades before a few biologists recognized that the two options were not all that different, and that they both showed that variation could exist within the natural processes of inheritance. Neither miracle nor mutation was necessary.

Mendel presented his earthshaking experiment to the forty members of his local natural history society. The Brunn (Brno) Natural History Society asked if they could publish the paper and he agreed. He sent out reprints to biologists he admired; unfortunately for him, one in particular, Karl Wilhelm von Nageli, a Swiss botanist, took the time to write back. Mendel was modest about his work but believed it significant, writing to Nageli, "I know that the results I obtained were not easily compatible with our contemporary knowledge and that under the circumstances publication of one such isolated experiment was doubly dangerous; dangerous for the experimenter and for the cause he represented." Indeed, agreed Nageli. Best to get more evidence. He recommended trying hawkweed, a plant that would have been disastrous for such an experiment because it often reproduces by parthenogenesis. Mendel would never have been able to reproduce his results with such a plant.

But in 1871 Mendel was made abbot, and running the monastery took up more and more of his time. A particularly virulent pea weevil invaded his pea plants. His efforts to collect hawkweed were thwarted by his being, in his words, "blessed with an excess of avoirdupois, which becomes very noticeable during long travels afoot, and as a consequence of the law of general gravitation, especially when climbing mountains." So Mendel did not publish again, nor did he send his results to more biologists. One or two mentioned his work, including a protégé of Darwin's, George Romanes, but no one much noticed it. Mendel died without knowing the immense effect his work would have, ultimately leading to a completely new field of biological science.

WHAT HAPPENED NEXT

Mendel's work should have given evolutionary theory a boost, revealing the mechanism for Darwin's variation, but instead it bogged things down. Now the theory faced a morass of contention made deeper by hard feelings. Two factions, one led by William Bateson and the other by Walter Weldon and his teacher, Karl Pearson, used Mendel's findings against each other, and their arguments quickly got personal. Their prime area of disagreement was over whether evolution was smooth and gradual or jumpy and discontinuous. They also disagreed in a fundamental way about each other's scientific methods.

Weldon investigated the evolution of crabs in Plymouth Sound, a highly polluted waterway. He noticed that crabs there had wider carapaces than the same species in other waterways, and he guessed this had to do with having a greater gill surface, making them better able to filter out pollutants. These crabs would be more fit, better able to survive and reproduce, and he set out to prove this in the lab. He and his wife collected thousands of crabs, raising them in five hundred large bottles filled with the sewage-rich waters of the Sound. He described the work as "horrible from the great quantity of decaying matter necessary to kill a healthy crab."

The smells may have been bad, but the numbers were good. Weldon was able to show that a wider carapace did result in a crab better able to survive in polluted water. He also used his results to assert that Darwin was right; evolution happened gradually, for the carapaces varied on a continuum—there were no suddenly larger carapaces. This work also allowed him to demonstrate his other conviction, that science was best when it was subjected to mathematical analysis. With the crab carapaces varying in a continuous fashion, differences had to be subjected to statistical tests. This was Weldon and Pearson's forte, as well as being what they saw as good science. Following in the footsteps of Darwin's cousin Galton, Weldon was sure that science would advance only as long as it was subjected to the rigorous application of math. Weldon was one of a class of biologists calling themselves biometricians.

Bateson, on the other hand, was math-phobic. He admitted to requiring extensive tutoring in math to qualify for his studies and then quickly forgetting it all. Statistical applications appalled him. Instead, he did his first work out of doors, traveling to the Central Asian steppes. There he studied marine invertebrates found in lakes derived from the Aral Sea. Each lake, conveniently enough, had different salt content, varying on a nearly continuous curve. The invertebrates, on the other hand, exhibited discontinuous variations. This supported his belief that the evolutionary change had to be saltatory.

Battle lines were drawn. Biology was to be, for the next few decades, split between gradualists and saltationists, biometricians and naturalists. In hindsight, much of what they fought over was not exclusionary. The dispute nonetheless slowed progress of evolutionary thought. This was due not a little to the vituperation between Bateson and Weldon. The two men were increasingly critical of each other, reviewing each other's work in more and more personal terms. They even got to the point of suppressing each other's publications.

The discovery of Mendel's work added fuel to the flames. Bateson read the paper after reading reports on experiments repeating the work by three Europeans (including de Vries) in corn, chickens, and mice. Somewhat incongruously, considering his distaste for mathematical methods, Bateson quickly accepted Mendel's conclusions. After all, to his mind, the particulate nature of Mendel's elements proved that evolution occurred in jumps.

Weldon did not agree, and he responded by attacking Mendel's experimental method. He protested that Mendel had never measured how green was green, and that pea plant seeds showed a continuous variation between yellow and green. Mendel's divisions were impossible to judge, he asserted, and it was too bad he hadn't made an attempt to describe them in a quantitatively accurate fashion.

Bateson struck back, defending Mendel and nonquantitative biology in general. Weldon and Pearson returned fire, criticizing Bateson's own experiments, finding his methods woefully imprecise. Of a paper by Bateson on crossing the hairy and nonhairy variants of the white campion flower, he asked, How hairy was hairy? He went on to bemoan imprecise methods in general, writing "the accumulation of records, in which results are massed together in ill-defined categories of variable and uncertain extent, can only result in harm."

All this general malignity came to a head in a meeting of the British Association of Zoology in 1904. Bateson had just been made president, but Weldon spoke first. Bateson replied. It was a rowdy debate. Pearson rose to call for a truce over the next three years so that the differences could be worked out. Finally, T. R. Stebbing, a cancer scientist, rose to bring the meeting to an end. A witness describes his speech as follows:

> In a preamble he deplored the feelings that had been aroused, and assured us that as a man of peace such controversy was little to his taste. We all began fidgeting at what promised to become a tame conclusion to so spirited a meeting, especially when he came to deal with Pearson's suggestion of a truce. But we need not have been anxious for the Rev. Mr. Stebbing had in him the makings of a first rate impresario. "You have all heard," said he, "what Professor Pearson has suggested" (pause), and then with a sudden rise of voice, "But what I say is: let them fight it out!"

THE SYNTHESIS

Weldon was not to live much longer, dying in 1906. Pearson kept the faith, struggling to keep Bateson honest, even if it took until 1921 before Bateson would finally admit that mathematical approach had merit. The feud over gradual versus saltatory evolution continued.

Science continued to pursue the study of evolution, but biology was on a path toward reductionism. Evolution meant different things to the scientists devoting themselves to deciphering inheritance from those studying development or paleontology. Within each discipline concepts began

111

to change, so that finally they weren't even talking the same language. Their solution was not to talk to each other much at all.

The greatest loser was natural selection. If there was one principle of Darwin's that was discarded—and to some it was good-bye and good riddance—it was Darwin's most essential, most innovative idea: natural selection.

GENETICS

With Bateson promoting Mendel's model, the study of inheritance began to take off. Under his influence, and with the influence of de Vries and his evening primroses, changes in organisms were assumed to be sudden. Very quickly the accepted mechanism for change from species to species, or even within one species, was assumed to be mutation.

American biologist Thomas Hunt Morgan visited de Vries's Amsterdam experimental garden in 1907. Morgan had trained in developmental biology, but was so impressed by de Vries's work that he changed his focus to inheritance. He was convinced by what he saw there that "Nature makes new species outright."

Even so, initially he was very skeptical about Mendelian theory, finding the whole idea arbitrary. He set out to see for himself. He started breeding mice, then rats. Then he hit the scientific equivalent of the jackpot; he tried flies, *Drosophila melanogaster*, aka the vinegar fly (or fruit fly or pomace fly or banana fly). He set up what was to become his famous fly room at Columbia University. The fly room was small, a mere 16 feet by 23 feet, stuffed with eight desks, the walls lined with shelves. Thousands of flies lived on those shelves in milk bottles, with electric light bulbs providing both light and heat for incubators. He fed them bananas.* The flies were the perfect subject—they were born, grew up, bred, and produced offspring in the short space of ten days. And they had only four chromosomes.

Even so, Morgan had an anxious and frustrating two years, waiting for a mutated fly. He finally got one, a famous white-eyed male spotted by his assistant. The next mutation was also a male fly, this time one with short wings. In all, he identified twenty-six mutations, all appearing in male flies— thus he called them sex-linked.†

As Morgan collected these mutated flies and bred them, he began to notice some odd things. Usually the mutant traits reappeared when the mutant flies were crossed, but not always. Morgan knew cell biologists had described a convoluted series of actions taken by chromosomes just pre-ceding a cell's division, called meiosis. In one step the paired chromosomes twisted around each other, finally breaking off and swapping whole segments. Morgan guessed that this crossing over could explain why some offspring didn't carry

---

* At least it was this way until improved by an assistant, Calvin Bridges, who worked out a feeding formula of agar, cornmeal, yeast, and molasses, and built actual incubators.

† This also located them on the short X chromosome, already associated with male sex by a previous researcher.

the mutant trait—the trait was swapped away during meiosis. They guessed that the farther apart the mutant traits were on the chromosome, the more often crossing over would result in their being separated, and not expressed. Using statistics, one of Morgan's students mapped the traits' likely positions on the chromosome. The closer together they were, the more often traits were seen in one offspring, which meant the traits were linked. By this time Morgan was a convert to Mendelian theory, and now he was busy refining it. His work was slowly turning the principles into a physical reality; not only could traits be mapped— suddenly you knew they were lined up, like beads on a string, and not just tossed in willy-nilly.

Back in Britain, Bateson was impressed by Morgan's work, although he admitted having a hard time reconciling the notion that inheritance had a physical basis with his current belief. Having a physical entity meant having a name, and the unit of inheritance began to be known as the gene (from Darwin's gemmules and de Vries's pangens). Bateson suggested a name for the new science: genetics.

Bateson made another suggestion. In his book *Mendel's Principles of Heredity*, he speculated that some human diseases could have an inherited basis. In particular he mentioned alkaptonuria, a condition creating discolored urine and skin and severe arthritis. Perhaps it was created by the combination of unfavorable recessive genes. This was read by a physician, Archibald E. Garrod, who just happened to be caring for an expecting couple, who just happened to be first cousins, and who both had siblings with the disease. When their child was born with alkaptonuria, he concluded Bateson's speculation was proved to be accurate. Garrod went on to theorize the disease was caused by a defect in a particular enzyme. He coined a phrase still used today for diseases representing simple genetic defects affecting protein formation: "inborn errors of metabolism."

In just fifty years science had gone from miracles to biochemistry. Such a change had profound implications; human beings depended on the same material mechanisms of life as animals and plants. That this finding had practical applications, promising to be of some help to mankind, may have softened the impact a bit.

The greatest practical applications were to come in agriculture. After World War I the Soviet Union suffered a severe famine. Much of the rest of the world was also facing food shortages, particularly of grain. Malthus's essay on population was coming true—population growth had exceeded the ability of the limited resources of land and agriculture to keep up. Lenin chose to emphasize prevention, taking some of the funds donated for famine relief to found the Institute of Applied Botany, dedicated to improving agricultural production. A talented geneticist, Nikolai Vavilov, was put in charge, and the institute made great strides when combined with research going on in the United States, producing new hybrid varieties of corn, vegetables, and livestock, all of which resulted in immense, almost exponential increases in yields.*

---

* These advances in agricultural production would prove, ironically, to undermine the very theory that triggered Darwinian thought. Malthus's theory of exponential population growth outstripping fixed resources was made obsolete, at least temporarily, by genetic technology.

Genetics became a hot field. Vast strides were being made in solving the problem Bateson described as obscure as lightning to a savage. The scientists working in genetics were growing confident they had the answers. To them, Darwin's ideas of gradual evolution and natural selection were as obsolete as special creation. They had seen with their own eyes what mutation could do.

Unfortunately, they had only a fuzzy idea of what mutation was. In fact, like de Vries, they called just about any change from parent to offspring a mutation. And they didn't think mutations were all that harmful. Morgan strongly believed mutations took a principal role, saying that if a mutation was not lethal, if a mutated form could survive and reproduce, the form would persist. If selection played a role at all, it was just for minor modifications.

Then, in the 1930s, de Vries's primroses suffered from serious scientific scrutiny. Three botanists repeated his work and discovered his new species were not species at all, but varieties. They weren't formed by mutation, either. Most were cases of chromosomal crossing over, some were cases of polyploidy (which happens when plants keep both sets of both parents' chromosomes). Debunking de Vries's mutants was a vindication for Darwinian evolutionists. We sense their relief in this quote by Cyril Darling, who said that de Vries's mutations "had given rise to a false theory of evolution, as well as a false theory of mutation and a false theory of heredity and chromosome behavior. Indeed, if we talk about scientific research blocking the progress of a science, Oenothera [primrose genus] was the imperishable example."

* * *

It was time for natural selection to make a comeback. Its first advocate was an unlikely one, Ronald A. Fisher. Fisher was a student of mechanical statistics at Cambridge, and turned to the statistical modeling of evolution because he had an intuition that Mendelian inheritance could be consistent with Darwin's natural selection.

Biometricians like Galton and Pearson had already worked on some models of inheritance. There was the Hardy-Weinberg equation, which calculated the frequencies that different genes, called alleles, will be found in a population. It showed that two alleles will persist at constant levels if not subjected to selection or any other outside influence. For an allele to persist in a population means that it is *fixed*.

Fisher envisioned genes more or less like the atoms in a gas, basing his model on Boyle's gas laws. He used the individual gene (or allele) as his unit, instead of an atom of gas. To fit this into a gaslike model he had to assume genes moved about freely and were distributed as uniformly in a population as are atoms in a gas. It also meant that in his model, forces like selection did not act on the whole organism, but instead on single alleles.

His goal was to find Darwinian variation in a model where inheritance worked on Mendelian principles. Fisher wanted to show that Mendelian principles by themselves would maintain variation in a population. He allowed for mutation, but built his model so that mutation was not the only creative force for evolution.

Fisher classified genes by their effect on variation. They would have an additive effect if they were recessive genes, and a nonadditive if they were dominant genes, which tend to make all organisms look alike. He kept the power of the genes in his model small; they didn't make large changes in the traits of his model organisms. He added two outside influences—selection and mutation—keeping the force of those effects small, too.

What he found was that a population would indeed retain the variation necessary for evolution, the variation that Darwinian theory required. Since Fisher was an admirer of Pearson, the premier biometrician at that time, he sent his paper to Pearson's journal, *Biometrika*. Pearson's reaction had more to do with reading the name Mendel (who, if you remember, his nemesis Bateson had been touting) than with the work. His comments on the paper were lukewarm, and since the other reviewer couldn't understand the math, the paper was rejected. It took two more years before Darwin's son Leonard pushed to have the paper read and published in 1918.

About this time, in the United States, Sewall Wright was doing his early work breeding guinea pigs. Coat colors and patterns in guinea pigs do not follow a simple Mendelian model—they tend to be inherited together. Guinea pigs with white collars tend to be certain colors and not others, and long-haired pigs tend to be other colors, for example. For Wright, genetics was not the simple reassorting of individual alleles. He began to understand it as the knitting together of genes in complexes.

Which is why, when he read Fisher's model, he wanted to improve it. Wright took a fundamentally different approach. Where Fisher took his inspiration from the regularity of gas laws, Wright took his from quantum mechanics and Heisenberg's Uncertainty Principle. Wright also had an opinion about another scientific argument of the time, one involving the movement of simple one-celled organisms. One opinion held that behavior at that level depended entirely on tropisms; the organism moved only when stimulated by a chemical, then only moving toward or away from it. The opposing opinion held that the organisms moved spontaneously and had the capacity to "learn" by trial and error. Wright was an advocate of the latter, preferring to give a role to spontaneity and random action versus passive tropism.

He gave the same role to spontaneity and uncertainty in the evolutionary model he designed. He changed the unit of his model—genes as he knew them did not act alone, they formed complexes. He focused on the whole organism, imagining it as the product of many gene interactions. Wright also knew populations were neither infinite nor uniform, and genes couldn't move about them as if they were. He saw natural populations as living and breeding in clumps—smaller populations that had occasional interactions with other clumps. He called these groups "demes"; they are also called effective populations. Effective populations are smaller than absolute populations because they include only reproductive individuals.

One of the results of Wright's modeling was uncovering the importance of population size. The results also meant a larger role for random events. The smaller the deme, the more likely a chance event will have an effect on it. For example, it is easier to believe that in a population of fifty animals, all offspring in one generation could be born male than it is to imagine it happening in a population of a thousand. It is equally more likely that in a small population one allele could just not make it into the next generation. Increase the population and you increase the chances that the allele will be reproduced, so it is far less likely the allele would be lost. Compare it to throwing dice—throw 1,000 times and you can be sure that at least once you will get snake eyes; throw them twenty times and it wouldn't be surprising if you didn't. Wright called this phenomenon "genetic drift."

Wright became just as famous for his adaptive landscapes. The diagrams are easy to understand and compelling (although the concept has been criticized as mathematically inconsistent). Adaptive landscapes look like topographical maps, with organisms existing on neighboring peaks. But the peaks do not represent real mountains; they represent summits of adaptation—populations exhibiting optimal adaptation to their environments. The valleys represent transitional stages—organisms that are less perfectly adapted. The problem posed by the landscapes is whether transitional organisms can change in time—that is, adapt quickly enough to make it to an adaptive peak—and not die out. The problem was made even more challenging by Wright's focus on the organism as his unit. This made evolution a bigger step; instead of Fisher's single gene having to change, a whole organism had to adapt.

Regardless of their disagreements, both models reestablished a role for natural selection in evolution. Fisher, Wright, and a third theoretician, J. B. S. Haldane, used their mathematical models to show that evolution was possible from the variation that was already present in populations. Mutation had a role, but not the central driving one envisioned by the geneticists.

In the meantime, an example of natural selection was happening right under their noses. What is most remarkable about industrial melanism, as it is called, is not that it took about fifty years to happen, but that it took fifty years to prove.

In 1848, a decade before Darwin published *Origin of Species,* a collector found the first-ever-recorded black or melanic form of the peppered moth, *Biston betularia,* near Manchester, England. Manchester had become a highly industrial city—the skies, buildings, and trees darkened by the soot of factories burning coal. In just forty-eight years, virtually all the peppered moths found in Manchester and other polluted areas were black.

It wasn't just happening to the peppered moths—some eighty other butterfly and moth species were adopting dark phases if they lived in polluted regions. The moths and butterflies in rural areas remained predominantly light colored. By 1915, geneticists had demonstrated that the trait was a simple Mendelian dominant trait. They held it up as an example of evolution happening in discontinuous, Mendelian fashion.

Others knew that couldn't be the entire answer. For one thing, simple gene dominance would not account for all the moths in a polluted area being black. An alternative suggestion was made that the change was due to mutations, brought on by the presence of pollutants. Perhaps the chemicals in

the air—sulfuric acid or metallic salts—were acting as muta-
gens, inducing genetic changes. This fit in well with a popular theory of the time that evolution progressed by means of
directed or "oriented" mutation (an updating of Lamarck's
idea of acquired change).

In 1924 Haldane made the point that such an explanation would require a mutation rate so high it would not be
likely, and suggested it was not parsimonious to propose it.
He held that a simpler explanation would be to suggest the
insects had been subjected to intense selection pressure.
Others had already made the case that the light moths were
easier to see on the darkened tree bark, and easier targets
for birds.

But biologists had for too long thought of evolution as a
slow change, and it was too hard for them to imagine that
such a simple thing as camouflage could make that much
difference. In studies published in 1937 and 1940, ecological geneticist E. B. Ford argued that the dark moths must
be physiologically advantaged, that they must be healthier
and more robust than the light moths.

Then, in 1955, biologist H. B. D. Kettlewell performed
the first proof of natural selection occurring in nature. He
captured, marked, and recaptured moths in order to closely
monitor their survival rates. Then he observed birds system-
atically, measuring the relative number of dark and light
moths they ate, comparing predation rates in rural and
industrial regions. He found the birds were taking two to
three times more light moths from the blackened trees in
Manchester, and two to three times more dark moths in

rural areas. Selection pressure had indeed been intense, and was all the explanation needed.

It is hard to say exactly when Darwin's natural selection re-emerged, becoming what's known as the *modern synthesis.* Selection made its way back in via the persistent arguments of people like Haldane and Ernst Mayr, and the reconciliation with genetics achieved by T. H. Huxley's grandson, Julian Huxley. The works recognized as turning points were Theodosius Dobzhansky's *Genetics and the Origin of Species* and Julian Huxley's *Evolution: The Modern Synthesis.*

There was one final component needed to make the synthesis complete. The presence of gaps in the fossil record had worried Darwin, and was used by detractors to reject his theory. Fossils rarely showed transitional forms (archaeopteryx was an exception) and seemed to show organisms persisting for long periods of time without changing. Darwin devoted chapters of *Origin* trying to explain away these discrepancies, arguing that the fossil record was inherently incomplete, that fossils were created only under unusual and special circumstances. While studies of genetics contributed to cementing the acceptance of evolution, paleontology lagged, finding no reason to accept gradual change. An American scientist, George Gaylord Simpson, brought paleontology into the synthesis. He set out to find a way to meld genetics and paleontology. He used the models of Fisher, Wright, and Haldane, drawing out their timelines to show, in his book *Tempo and Mode in Evolution,* that the fossil record was indeed consistent with the new model of Darwinism.

Darwin's theory of natural selection had re-earned its legitimacy, but his belief in the gradual nature of evolution still faced challenges. Speciation, when an organism changed from one species to another, was particularly hard for many to imagine as a gradual event. Other theories were proposed, although the role of mutations was less emphasized. Geneticist Richard Goldschmidt suggested a mechanism wherein the reorganization of the chromosome would lead to the creation of what he called "hopeful monsters." He did not support single mutations as the source of change, and he was not willing to accept the "beads on a string" theory of chromosome structure. Instead, he imagined the chromosomes as arranged in a sort of molecular field, which could instantly and radically change, leading to significant changes in the resulting organism. Like mutants, most of the "monsters" would be poorly adapted and die, but some might have advantages. He gave as an example the manx cat, a breed of cat born without a tail or with a shortened, twisted one. This provided the manx with no known benefit, but perhaps it would lead to the sort of change that gave rise to birds. His example was the short, fanned tail of the transitional archaeopteryx.

Goldschmidt did not go so far as to postulate a precise mechanism for this change. His ideas were ultimately dismissed, mostly because theorists like Mayr and Huxley were able to demonstrate more and more convincingly that gradual change could explain what was observed in nature.

So the case for the synthesis between selection and genetics was made more and more convincingly. Then came a setback, a paper by Herman J. Muller. Muller had been a student of Morgan's, then was recruited by Vavilov to bring the fly room to Moscow. He worked there for seven years, finally leaving when the Soviet attitude about the study of genetics changed significantly.* His experiments included the early use of X rays to induce mutations. He exposed fly chromosomes to high doses of X rays and obtained mutations at up to 150 times the normal rate. He used these mutations to help him map over 500 genes on *Drosophila* chromosomes.

After returning to the United States in the 1950s, he published a paper proposing a method of calculating the "load" of mutated genes present in humans. For his calculations he made the assumption that, other than a few gene sites (by his estimates, between eight and eighty), all genes in humans were homozygous—meaning the genes coded for the same traits. He believed that in those few places where genes might vary, maybe one out of a thousand, the alternative copy would most likely be recessive and lethal. This fit the prevailing opinion of most experimental geneticists. It was the assumption held by both Fisher and Wright in their early models. It reflected the traditional view, older than Darwin, that species were basically pure. If mutations

---

* Soviet science changed dramatically under Joseph Stalin. At first the Russians almost deified Darwin, placing him on a par with Lenin. In part this was because of his atheism, and in part because his theory jibed with a merit-based society. When Stalin took over, he appointed Trofim Denisovich Lysenko head of research. Lysenko believed that Lamarck's theory of acquired traits fit better with the idealism of the communist system. More about this in chapter 6.

occurred, they were either quickly eliminated (if they were deleterious) or just as quickly assimilated into the homozygous state (if beneficial).

This was in direct opposition to the view held by naturalists. While their ideas took a back seat to the more glamorous advances made by the experimental geneticists, some of their work was being recognized. In particular, in the United States Theodosius Dobzhansky, author of one of the most important texts on the synthesis, strongly believed heterozygosity was a natural condition for wild species. He learned most of his genetics in Russia, from Sergei Chetverikov. Chetverikov, like most Russians in the early years of the century, was deeply affected by Darwin. These Russian scientists had never doubted natural selection as an important force guiding evolution. Chetverikov had read the work of the American fly geneticists and had set out to test their conclusions using wild fly populations. He was equally interested in proving that enough variation existed in wild populations to support Darwinian natural selection. He captured 237 female *Drosophila* from around Moscow, then brother-sister mated their offspring. In one generation he found thirty-seven hidden traits—far more variation than was suspected by other biologists.

This work was not much read in the west, not a little because it was seldom translated. Dobzhansky's emigration (a mirror image of Muller's trip), and his popularization of Chetverikov's conclusions, went a long way toward bringing this work to the attention of U.S. researchers. Chetverikov is an unsung genius; some of his models were more advanced than those of Fisher or Wright.

Chetverikov's models, as well as Dobzhansky's, assumed the presence of considerable genetic variation in the wild. If Muller was right, and the natural state of species was one of predominant homozygosity, it would make most of population genetics immaterial and wrong. Dobzhansky was sure Muller was wrong and set out to prove it.

This disagreement was not purely scientific. Muller was a strong believer in eugenics—he believed that the human species could be improved only through artificial selection. If humans were mostly genetically homozygous, it meant that natural selection and evolution for the species had mostly ended. Dobzhansky was passionately opposed to eugenics. If he could prove that humans carried a great deal more variation than Muller thought, it would mean evolution on a natural basis was occurring. In addition, if the natural state was one of considerable variation, artificial selection could have little predictable effect.

Dobzhansky quickly recruited other scientists to help collect experimental evidence for the natural genetic variation of species. Initially they attempted breeding experiments to measure variation, but the results were fraught with problems. For the most part it was impossible to make the data statistically significant.

Real proof had to come from the invention of a new experimental technique, protein electrophoresis. In those days proteins were sort of large molecular black boxes. There were a couple of things that could be tested about them—their size and their electric charge. Experimenters could render them into a partially solidified solution—a filmlike gel—pass a current through the gel, and watch as the proteins would slowly migrate from one electrically charged pole toward the other. Turn off the voltage and they would stop. How fast and how far each protein moved would depend on its size and the amount of charge it contained. Even small differences in its amino acid sequence would result in its ending up in slightly different places in the gel.

Since proteins are the end result of genes, it was a simple and conclusive test. Mash up any organism, extract the proteins, put them into solution, turn on the current, and voila: electrophoresis. If the proteins lined up differently, they were different. If they were different, so were the genes that made them.

Two research groups, one in England led by H. Harris, and one in the United States led by Richard Lewontin and J. L. Hubby, put proteins through electrophoresis. Since this was still a crude tool (it is quite conceivable that proteins could be the same size, carry the same charge, and yet still be very different), the results would underestimate the actual differences. Still, they showed that about one third of the genes of a species were protein polymorphic—not the same. As more species were tested, the same results were obtained over and over. As tests have become more sophisticated—as protein sequencing and gene sequencing have become possible—the results continue to be borne out. The genes carried by any species are predominantly heterozygous—every species carries a great deal of genetic variability.

These results forced geneticists to think in an entirely new way. Instead of thinking of genes as determining the organism's outward appearance, or phenotype, they had to readjust, and think of the interplay of genes as the formative source. A cat looked like a cat not because the genes dictated what the cat should look like, but because the various actions of the genes, the balance achieved by these actions, did. Genetics was taking a step beyond Mendel.

6

Impact

One mild surprise nestled in the history of evolution is that "survival of the fittest" isn't Darwin's phrase—in fact, he didn't much like it. He did, grudgingly, include it in his fifth edition of *Origin of Species,* because it did have a certain appropriateness and precise pithiness. But he had his reservations.

The phrase has been a burden ever since, becoming the battle cry for the darkest interpretation of Darwin's idea: social Darwinism. It is often treated as if it *were* the theory. Unfortunately, the phrase is a tautology. Start with the question: Who or what is the fittest? Answer: those that survive. Question: Who or what survives? Answer: the most fit. One mode of attack on evolutionary theory holds that if the phrase is a fallacy, so, too, is the theory.

So Darwin's reservations were on the mark. The phrase was Herbert Spencer's. You may remember that Spencer wrote a theory of evolution a few years before Darwin, but since he used it to explain all of life and culture, Darwin didn't find it upstaged his own work. Spencer was not a biologist and gave no theories for the mechanisms of evolution. Plus, he echoed the Greeks in using a linear model—his own version of the great chain of being. His friend T. H. Huxley advised him he couldn't explain anything in biology as a chain, you may recall, telling him the organization was more like a tree.

Still, Spencer was a respected writer and philosopher. Darwin thought him perhaps the greatest living English philosopher, although he also admitted to being able to understand only some of his writings. Spencer's ideas are still taught; however, his reputation has been sullied, and sullied evolutionary theory by association.

Spencer, the poor son of a schoolteacher, trained as a civil engineer for the railways. He abandoned this profession and managed to make a reasonable living in journalism and philosophy, a success that helped make him Huxley's friend and mentor. Like Huxley, he held strong opinions and was eager to voice them. One failing he had was that he read little, and refused to read writings by those he disagreed with, saying it would do him no good.

His ideas arose out of some of this disagreeableness. His core premise for biological evolution was inspired by reading Charles Lyell's refutation of Lamarck at the end of his *Principles of Geology*. He remained a supporter of Lamarckian theory to the end of his life, although he also supported Darwin's theory.

Spencer pointed out that the vast numbers of species, about ten million by contemporary estimates, could hardly be expected to have all been specially created. Instead he theorized they were all the result of "continual modifica-tions, due to change of circumstances." He used biological evolution as an analogy for the larger picture. If organisms were subject to continual tinkering, so were societies and cultures, so must be the world. And there must be rules, laws, guiding the continual change: "when we regard the differ-ent existences with which they severally deal as component parts of one cosmos; we see at once that here are not sev-eral kinds of evolution having certain traits in common, but one evolution going on everywhere after the same manner." This evolution had a direction, too. For Spencer there was some sort of progress toward increased complexity.

Spencer approved of Darwin's ideas, which proved unfor-tunate for evolutionary theory. In fact, once Darwin pub-lished, Spencer applauded and then co-opted his work, using the principle of natural selection in his own writings. In this way Spencer was the originator of social Darwinism.

In Spencer's view, what is good for nature is good for man. Mankind and society would be better off if left to struggle the way organisms in nature must. The struggle would improve those who survived, and remove those who could not. Social welfare programs were counterproductive, because they were aimed at promoting survival for the poor, mentally retarded, and other imperfect people. He wrote, "If they are sufficiently complete to live, they do live, and it is well they should live. If they are not sufficiently complete to live, they die, and it is best they should die."

This laissez-faire sociology was accepted and used as a justification for many social injustices. It helped justify colo-nialism and an imbalance in the distribution of wealth.

Imperialistic nations set out to improve "undeveloped" cultures. Very wealthy capitalists in the United States embraced Spencer's work as proof of their own worthiness and entitlement. Upper-class and middle-class Britons also found it easy to believe that they deserved their relatively comfortable fates. The dismal lot of less fortunate people, such as Irish famine victims, could be regarded as just in that it served to remove the less fit; regrettable, perhaps, but correct, and above all *natural*.

Another bit of fallout from Spencer's promotion of survival of the fittest was an emphasis on struggle. This was also highlighted in many interpretations of Darwin's writings. Christians would bemoan the fact that evolution and natural selection assumed a world always in conflict. To imagine such a world was to their minds depressing and dispiriting. They usually interpreted the struggle to be a direct one between individuals, or, more warlike, between species. They reasoned that this was dangerous for the future. To describe conflict as a natural state was dangerous, for it condoned conflict and gave validity to war.

Eugenics

A somewhat more scientific aspect of social Darwinism was eugenics. Spencer himself persisted in his belief in acquired Lamarckian improvements, which told him that poor people who worked hard and did improve would have improved children, made better by their parents' efforts both in condition and inheritance. As it became more clear at the turn of the century that Weismann, the cytologist who wrote of the continuity of germ plasm and chopped off mice tails, was right, and there was no acquired inheritance, whatever benign nature there was to Spencer's social prescriptions began to fade.

The result was eugenics, a term coined by Darwin's biometrically inclined cousin Francis Galton. The word means simply "good birth," but it led to evil.

Galton was interested in the inheritance of intelligence and talent. He noticed that biological genius ran in his family (Erasmus and Charles Darwin) and musical genius and legal genius ran in other families. He concluded that there must be an inherited basis for these traits and set about to calculate formulas to predict their transmission. One of his conclusions—that traits are passed on in declining fractions from parents, grandparents, and so on—was challenged by Mendelian genetics.

Galton also started a eugenics society and established eugenics as a legitimate subject for scientific study. Many scientists and social scientists took it up. Galton himself believed that the human race and society could only benefit by the knowledge he was seeking. He advocated only positive measures be taken to encourage "good" families to have many children—for example, family allowances to encourage and enable better citizens to produce large families.

Darwin himself agreed with most of his cousin's conclusions. He was quite concerned with his own family, believing his children suffered the ills of inbreeding; his having

married his first cousin became a great source of guilt for
him.

In the meantime, eugenics was being taken up by scientists in other countries. In Germany, one of Darwin's
strongest, most outspoken supporters, Ernst Haeckel, was
coming to some dangerous conclusions. A gifted biologist
and embryologist, Haeckel is nonetheless better known for
some seriously wrong-headed conclusions. He was also a
clever phrase-maker, inventing the word *ecology* (for "household of nature") but was best known in biology for his
phrase "ontogeny recapitulates phylogeny."

What this phrase says is that as an embryo develops, it
passes through stages representing lower organisms. So both
avian and mammalian embryos grow gill slits, although they
will never use gills, and humans have tails during part of
their potential development. From this Haeckel derived
what he called his biogenetic law.

This biogenetic law is really just a restating of Aristotle's
great chain of being. Haeckel basically was dragging biology
back to the concept of progress, and a ladder of development, leading to the greatest organism of all, man. (And
not woman—Haeckel placed white women down a rung or
two, alongside white infants, Negroes, and "the simious
type.") This biogenetic ladder was extended to species and
human races and cultures as well, with the white northern
European race and culture being the paragon of evolutionary achievement. Not even religions evaded his grading; he classified Judaism as a sort of embryonic Christianity. All other races and cultures were in a state of arrested
development.

Haeckel's biogenetic law had already been disproved, even before he wrote it. A countryman, Karl Ernst von Baer, had already observed that embryos share common stages, but that no embryo ever represents a mature stage. In effect he was saying that embryos may pass through stages similar to the embryos of other species—thus all embryos have gill slits—but there is no *progression*, just common stages; all species become more specialized as they approach their adult stages. In this way Baer helped develop the branching concept of species, but he never did accept evolution, even though he died long after the publication of *Origin*.

Still, Haeckel's ideas were widely accepted and popular, particularly among white European males. He devalued the lives of "lower" races (including Jews). He claimed that Jesus must not have been Jewish, because his thinking was too advanced. Instead, Haeckel reasoned, he was probably the son of a Roman soldier (of German descent, no less). Not surprisingly, Hitler adopted Haeckel's biological recommendations, paraphrasing Haeckel in his discussions of race in *Mein Kampf*.

Germany wasn't the only nation embracing this new "science" and using it to inform their policies. In England it remained relatively benign, merely something the government quietly encouraged. The United States selected desirable and undesirable immigrants based on what were perceived to be their potential genetic contributions. The United States also practiced selective sterilization, as did Sweden and Denmark. Eugenics as a policy crossed party lines, being favored both by the far right (Nazis) and socialists, among whom we can count Karl Pearson of England

and American Herman J. Muller. Only countries dominated by Catholicism rejected eugenics outright.

In Sweden, a politician who later admitted to being a Nazi promoted eugenics as government policy. Even though he was ousted, eugenics was not; instead, the next leader, a social Democrat, promoted it as a public good. Funding went into the laboratory study of genetic disease. Even as the government leaned farther left and into socialism, the government initiated sterilization of people who were considered inferior. The academics who promoted these policies criticized other uses of eugenics as racially biased and class based, but they still picked up itinerants and "different" individuals for involuntary sterilization.

Muller, Morgan's assistant who took the fly room to the Soviet Union, was a eugenicist. He signed a "Geneticists' Manifesto" promoting eugenics for health reasons, but also promised the practice could promote the birth of intelligent humans imbued with something he called "success." This trait was tailored to socialist and communist societies; it meant having behavior reflecting a sense of social responsibility, being a "comrade," in a sense. Muller believed this relatively gentle application of eugenics would be ultimately beneficial, and that a day would come when "everyone might look upon 'genius' . . . as his birthright."

In the hands of the United States government, eugenics took a more oppressive tone. In several states involuntary sterilization was practiced on women considered to be unfit. Any poor, criminal, or drunk citizen could earn this distinction. There were also strong immigration biases, with immigrants from southern European countries as well as Asian and African nationalities discouraged from settling, if not banned outright. In addition, states passed laws against interracial marriage, reflecting the belief that children produced of such pairings would be inferior.

Eugenics held the public's attention, as well. People enthusiastically accepted eugenics in a pseudoscientific form. The best-known eugenics expert was Charles Davenport, a one-time chicken breeder. He assigned genetic traits to everything human, inventing terms like *thalasophilia,* or love of the sea, as necessary to make a sea captain. Thalasophilia was even classified as a sex-linked trait because, after all, sea captains were pretty much exclusively men.

Eugenics was put in service of morality plays, in particular with the case of the Jukes and the Kallikaks. These two families were held up as the ideal bad example of how human breeding can go wrong. The Jukes descended from a Dutch immigrant to New York. The first Juke was described as feebleminded and dishonest, and just about all of his descendants were also feebleminded, dishonest, and disreputable—a perfect case of genetically determined degeneracy. Even more convincing was a second case of two families, given the pseudonym Kallikak (taken from two Greek words, one for good and one for bad). The story started with a revolutionary war soldier, who during his training became involved with a prostitute, subsequently fathering a child by her. After his service in the war, in which he was honorable and heroic, he essentially abandoned the woman and had nothing to do with her or his son. Instead he started life anew and married an equally honorable and respectable woman. When a sociologist later traced the descendants of each family he found the first family produced nothing but thieves and prostitutes. In contrast, the second family produced nothing but lawyers, judges, and doctors. Since both families lived in the same environment— they both lived in Pennsylvania—the difference had to be genetic. Just a small amount of degenerate effect then, could have devastating results.

Eugenics even took on an agricultural mien in public demonstrations. County fairs held "fitter family" competitions, where families would be shown off along with the best hogs and pumpkins, their blond hair and pink faces shining.

Of course the most horrible extension of eugenics was the Holocaust. Hitler started his genetic cleanup with not only sterilization but euthanasia of the mentally retarded. Shortly the classification of undesirables and degenerates expanded to include people of non-Aryan races: Jews, Gypsies, and those living "nonconventional" lives, like homosexuals and Catholics.

It is easy today, with our improved understanding of just how complex the workings of heredity are, to condemn eugenics as not just cruel, but also ridiculous. But genetics was then a new science. In a way, the accumulation of new knowledge about genes outstripped the understanding. There were some very basic deficits in that understanding; for example, scientists still believed organisms were the products of relatively pure genes. Homozygosity was still

believed to be the status quo. They were also still thinking in terms of adaptations where every change and difference was adaptive. Random chance had yet to make its entrance; nature was still seen as regulated and balanced and
not as the chaotic melee, stuffed with opportunity, that it
really is.

But social Darwinism did its work. The science of evolution gained a dark pall it hasn't been able to shake to this
day. Despite the words of scientists like Muller, who, even
while they supported eugenics as a science, also recognized
the tendency for it to be used for entirely racist purposes,
Darwinism seemed allied with social evil.

And it was not just because of eugenics. Christians
hated the premise of natural selection based on competition, which gave the theory an emphasis seemingly based
on struggle and strife. While religion spoke to charity and
predicted the meek would inherit the earth, evolution
seemed to suggest the weak should just die. There seemed
to be a fascination with aggression and its role in behavior
and selection that persisted into the second half of the
twentieth century.

Yet even as long ago as 1902, the concept that cooperation between organisms held a role in selection was being
discussed. A Russian philosopher, Peter Kropotkin, pointed
out that cooperation was common between animals. Kropotkin's book *Mutual Aid: A Factor of Evolution* was a study of
altruism and aid in animal societies and discussed the selective forces that might create such characteristics. Unfortunately, Kropotkin's book gained little attention.

CREATIONISM

Given the relatively brutal associations of social Darwinism, it's no surprise that a backlash ensued, particularly in the United States. This backlash persists to this day in the form of the Creationist movement, and also exists within other social movements like feminism and multiculturalism. Social Darwinism in other guises persists, as well.

In the United States, fundamentalist Christians were quick to reject evolutionary theory, and with it social Darwinism, and were not willing to make any distinction between the two. They worked hard to effect bans on teaching evolution in public schools or including the theory in textbooks. They equated belief in evolution with antipatriotism, claiming it contrary to the principles of the Founding Fathers.

This movement crystallized in the Scopes trial, also known as the Monkey Trial. Held in Dayton, Tennessee, in July 1925, the trial became a media event, commanding as much national attention in its day as the O. J. Simpson murder trial in the mid-1990s.

The Monkey Trial was *intended* to be a public spectacle. In response to fundamentalist pressures on state and local governments to restrict teaching evolution, the American Civil Liberties Union (ACLU) was in active search of a test case. A mining engineer in Dayton, George Rappleyea, knew of this search. He and his friend John Thomas Scopes, a local teacher, agreed that science, particularly biology, could not be taught without including evolution. Rappleyea convinced Scopes he should volunteer to be the test case.

Scopes stepped up, and they went to local authorities; Scopes admitted he was breaking the law and teaching evolution.

The town itself welcomed the attention. Town officials knew it would bring the national media. The larger neighboring town of Knoxville even tried to horn in on it, attempting to change the trial's venue. Dayton, they suggested, wasn't big enough to put on such a trial.

They were right. The courthouse could seat 700 people and generally 300 more people squeezed their way in. On the last day, the trial was held outside; officials were afraid the heat and the tight confines of the courthouse would result in heat-related casualties.

The trial met the public's expectations. First, two famous lawyers volunteered to argue each side. Clarence Darrow, a legendary criminal defense lawyer, agreed to stand up for the ACLU and John Scopes. He was also famous for being an agnostic. His opponent would be William Jennings Bryan, a spokesperson for the fundamentalist movement and perennial presidential candidate. Neither lawyer seemed likely to hold back. Bryan described the trial to the media as "a duel to the death" between evolution and Christianity.

Exaggeration was a theme. Journalist H. L. Mencken called the trial a circus, and somewhat rightfully so. Outside the courtroom, the town attracted circus chimpanzees, Bible thumpers, vendors selling monkey dolls, and a new drink called monkey fizz.

Darrow for his part tried to make the point it was not Christianity per se but fundamentalist Christianity that insisted on taking the Bible literally. Darrow was prevented from doing this by the judge, who refused to allow most of Darrow's expert witnesses, either sending the jury away during their testimony or only allowing written statements. Darrow contributed his own bit to the theatrical atmosphere by protesting the judge's starting every session with a prayer, and asking that the sign over the judges bench admonishing all to "Read Your Bible" be covered or taken down.

Thus blocked, Darrow, in a legendary move, called Bryan to the stand. The judge protested this, but Bryan was only too willing. Darrow then set out to reveal that not only did Bryan not know his Bible, but that even he, the representative for fundamentalists, did not take every word literally.

Mr. Bryan, do you believe the first woman was Eve?

Yes.

Do you believe she was literally made out of Adam's rib?

I do.

Did you ever discover where Cain got his wife?

No sir. I'll leave the agnostics to hunt for her.

Does the statement "The morning and the evening were the first day" and "The morning and the evening were the second day" mean anything to you?

I do not think it necessarily means a 24-hour day. My impression is that they were periods.

Have you any idea of the length of the periods?

No, I don't.

Do you think the sun was made on the fourth day?

Yes.

And they had evening and morning without the sun?

I am simply saying it is a period.

They had evening and morning for four periods without the sun, do you think?

I believe in creation as there told, and if I am not able to explain it, I will accept it.

Other parts of the transcripts held little more than sniping and grandstanding. For example, asked what the relevance of the questioning was, Bryan volunteered: "The purpose is to cast ridicule on everybody who believes in the Bible, and I am perfectly willing that the world shall know that these gentlemen have no other purpose than ridiculing every Christian who believes in the Bible." Darrow snapped back, "We have the purpose of preventing bigots and ignoramuses from controlling the education of the United States and you know it, that is all."

Most interpretations of the trial describe this examination by Darrow of Bryan as being a defeat for Bryan. Even evangelical papers admitted Bryan had not defended the Bible as well as could be expected. Still, the jury found Scopes guilty and fined him $100, the minimum fine. Scopes went on to study geology at the University of Chicago on a scholarship provided him by grateful scientists. Bryan died of complications of diabetes just days after the trial, his death represented variously as being the result of his defeat and humiliation at the hands of Darrow. After the Scopes trial, Tennessee, Mississippi, and Arkansas kept their anti-evolution laws on the books but did not actively enforce them.

William Jennings Bryan tends to be represented (as in the play and movie *Inherit the Wind*) as an intolerant blowhard. He undoubtedly enjoyed the sound of his own oration, but his motives were not purely oppressive. If one can get past the religious aspects of his rejection of Darwinism (where the struggle going to the fittest contradicts the Christian edict favoring the meek), the social merit of his stance is obvious. Bryan was known as the "Great Commoner," and thought of himself as a defender of the common man. (Indeed, it bothered him to be the prosecutor in this case; to attack a man like Scopes went against his principles. He preferred to think of himself as defending the masses against the moral degradation of Darwinism.) He rejected the inequities of class and race built into social Darwinism. He simply could not separate out the scientific merit of the theory of Darwin from the societal misuses of social Darwinism.

Bryan was a crusader and spokesperson for Christian fundamentalism, a movement just getting started at the turn of the century. The name, fundamentalist, came from pamphlets distributed in Chicago. One pamphlet offered an interpretation of Darwin's theory that could accommodate the scriptures. It described Darwin's theory of life in terms of God having created a few forms with an enormous capacity for variation. In this way the origins of fundamentalism were not as anti-Darwin as later fundamentalists have become. (It is also interesting to note that besides Darwin's theory, the practice of biblical criticism, a liberal approach to religion that revealed inconsistencies in the Bible, also came to the United States at about the turn of the twentieth century.)

Creation science arose as efforts to ban the teaching of evolutionary theory were thwarted by the courts. It is not only a fundamentalist movement, and boasts representatives from Christian and Judaic and Muslim faiths as well. It did not die with the Scopes trial; indeed, the trial may be better described as its birthplace.

Contrary to the tale told in *Inherit the Wind*, state governments did not immediately revoke their "monkey laws." The conviction of Scopes was condemned; Bryan was considered embarrassed (and his subsequent death validated the image of him as humiliated and broken). Still, governments were still feeling the pressure from creationist constituents. The Tennessee law stayed on the books until 1967, and so did similar laws in Arkansas and Mississippi.

With the public humiliation of the Scopes trial, the creationists tried new strategies. The movement went local. They became school board members, and influenced curriculums from there, and they pressured textbook writers and even harassed schoolteachers. Evolution began to sink below the radar screen of American classrooms.

It took the space program of the 1960s to make evolutionary theory legitimate again in the United States. When the Soviet Union put a man into space, suddenly eclipsing U.S. accomplishments, the need for science education became all-important, a major political issue. Evolutionary theory then piggybacked on the drive to advance students' understanding of physics and mathematics, so the United States could compete in technological achievement.

Infused with a new confidence, science teachers challenged the old laws. In 1968 a schoolteacher from Arkansas won a decision from the U.S. Supreme Court, which found that teaching creation on an equal basis with evolution was unconstitutional. The Supreme Court used as its justification the constitutional mandate to keep religion separate from government. It was clear, Federal Judge William R. Overton wrote, "Both the concepts and wording . . . convey an inescapable religiosity," and that "no group, no matter how large or small, may use the government . . . to foist its religious beliefs on others."

Creationists tried a new tack. Tennessee passed a replacement act in 1973 making it mandatory to teach Genesis as an alternative theory to any discussions of creation or origins. Federal courts quickly struck down this law, also on the basis of the separation of church and state. A similar attempt was made in Arkansas in 1981, with the same result. Fundamentalists were unbowed. The representative who sponsored the Arkansas bill, after the courts struck it down, remarked, "If the law is unconstitutional, it'll be because of something in the language that's wrong. . . . So we'll just change the wording and try again with another bill. We got a lot of time. Eventually we'll get one that is constitutional."

It was clear they had to promote Genesis as something other than a religious doctrine. Their solution was to give it a scientific spin. They approached this using two strategies. One was to provide scientific evidence for Genesis (finding evidence that seemed to point to a flood having resulted in deposition of fossils). The other was to collect any bit of evidence that appeared to contradict Darwinian theory. Creationists jumped at any sign of scientific quibbling. If a biologist questions any part of current evolutionary thought, creationist scientists have their banner headline. Renowned paleontologist Stephen Jay Gould and Niles Eldredge's theory of punctuated equilibrium famously questioned the accepted mechanisms of gradual speciation and instantly the two were represented by creationists as scientists who no longer believed in Darwin's theory. Both were quick to publicly deny that their comments had had anything to do with a reversion to belief in special creation.

Creation science itself is far from uniform—various groups believe in various versions of Genesis. Some will accept that humans have descended from apes; others match paleontological periods to the days of creation in Genesis. Finally there are those who still adhere to a literal translation of biblical scriptures—that the universe was created six thousand years ago in six twenty-four-hour days and every organism was created individually. They all have one thing in common: they wish to at least have their ideas presented alongside evolutionary theory in the education of schoolchildren. Some, if not all, would like Darwinian evolution removed completely from the school curriculum.

Creation science and fundamentalists have had some success. Currently more than half the American public attest to believing in some sort of Genesis-based version of the origin of life on earth. Even more believe that alternative versions of origins should be taught alongside evolutionary theory. This should not be a surprise, since many of these people grew up with textbooks that barely mention Charles Darwin, if they mention him at all. One did refer to him—as the famous father of George Darwin.

LYSENKOISM

Creation science inspires a cautionary note from some scientists and philosophers: not only do we need to fear the influence of religion on government, we need also to fear the influence of government on science. There are two very good examples: Nazism and Lysenkoism.

Trofim Denisovich Lysenko was the "dictator" biologist of the Soviet Union from 1948 until his ouster by Khrushchev in 1953 (he remained a scientific advisor). He became the preeminent scientist less by talent than by fitting his scientific ideas into the ideology of the Soviet government, which at that time meant Joseph Stalin.

Lysenko fit the communist ideal of a scientist, something not common to scientists of the time. He was famously called the "barefoot scientist," highlighting his peasant origins. He was a true proletarian, not of bourgeois origins as were better known and established biologists, such as Nikolai Vavilov.

Vavilov was also a dedicated socialist who believed that a socialist system inherently offered greater freedom for scientific practice than the capitalist and theocratic governments of the West. Supported by Lenin, he was influential in putting his ideals into action, establishing institutes of agriculture and genetics that rivaled any in the world. He was able to persuade H. J. Muller, who had socialistic leanings, to leave the United States and work in Moscow even after he had won the Nobel Prize for his work describing mutations caused by radiation. On the other hand, Vavilov complained of Lysenko's lack of culture as well as his scientific ignorance. He also insisted on studying Mendelian genetics and neo-Darwinism when Stalin preferred neo-Lamarckian theory. The idea that will power could improve a species (or a people) fit in with Stalin's plans. He would make the Soviet people into communists, through acquired characteristics.

Lysenko's politics were impeccable, inspiring bravos from Stalin. His science was not. His claim to fame was a process called vernalization, in which kernels of spring wheat are soaked and then refrigerated. The result is that they flower sooner and can be grown as winter wheat. That this was a common practice invented by someone else didn't inhibit Lysenko, even when the inventor protested publicly— Lysenko claimed it as his own. Not only that, he claimed the changes were then permanent (as acquired inheritance) and so the process didn't have to be repeated. They're not, and it does.

Once installed in 1948, he imposed many equally unhelpful ideas on Soviet agriculture, such as cluster planting, and planting potatoes in the middle of the summer. The results were often disastrous, leading to famine. His scientific method was equally arbitrary. He refused to apply mathematics to his science, because once others had found a mistake in his calculations. He did experiments to get the results he wanted, limiting numbers, and often omitting controls to fit his purpose. Under his control genetics was removed from all scientific curricula, even medical study. In his words, "genetics is merely an amusement, like chess or football."

Worst of all, he did what he needed to do to squelch his competition. Many scientists were exiled or simply vanished;

Muller left in protest and returned to the United States. Vavilov resisted loudly, arguing that Lysenko's ideas were outmoded. In response Vavilov was officially criticized, and removed from all his posts. He was finally tried and found guilty of sabotage in agriculture and faced the death penalty. His sentence was commuted, but he died in a gulag in Siberia.

Lysenko took over a nation's science because he sided with government-supported ideology. Under his control, all Soviet science was performed within an ideological framework. As a result, biological science and agriculture in the Soviet Union stagnated. People starved. A small number of scientists in the United States today warn that something similar could happen if creationists ever succeed in having creation science taught alongside evolutionary theory.

Back in the lab and the field, scientists at mid-century paid scant attention to how society perceived their endeavors. They were too busy. Evolution was taking on new hues with the unveiling of the gene as a molecule called DNA. At about the same time, game theory and chaos made inroads into explaining the unexplainable about natural selection. Evolution remained controversial, a science of opposites. But it had gained a sophistication and viability taking it far beyond the quibbling of creationists.

7

Post-Normal Science

If you are lucky enough to travel to the Galápagos Islands, and you know nothing about Darwin or the theory of evolution, you might wonder what all the fuss is about. You will find only the same barren, unproductive islands that Darwin first described. "Nothing could be less inviting than the first appearance," he wrote. "A broken field of black basaltic lava is every where covered by stunted brushwood, which shows little signs of life."

If you go there with Darwin as your guide, however, the islands will come alive. The parched, brushwood-covered hills will be as sparse and uniform as Darwin described, but you will know how far the plants traveled. The creatures will be as hard to find and as dully black or gray, but you will see each one's unique adaptations. The bright boobies and swimming iguanas will thrill you as they do the unprepared, but you will know more of their secrets. You will see each with the sharp eye of science—each tiny finch's beak will matter, each lava lizard, each mockingbird a specimen to compare from island to island. For the first time in your life, perhaps, the tiny and dull will catch your eye and your breath; you will strain to see every detail.

You will look so hard at the tiny birds that when you look up you will be surprised. For the island supporting you is new, and it shows. The lava is fresh and still black and it stretches for miles in all directions. The surrounding seas are cool and flush with energy—they run currents as fast as rivers and teem with life. And you will feel how far away you are from anywhere else in time and in distance.

That is when the Galápagos will emerge for you. The place is bigger than it is, greater than its component parts. It is more than a site for volcanoes and quixotic creatures. It is a place far from equilibrium; it is a place in evolution. When you see this, you begin to understand how normal science shortchanges us. There are mysteries it doesn't have the patience to unravel, and for anyone touched by the Galápagos those are the very mysteries that matter.

POST-NORMAL SCIENCE

Will evolutionary theory redefine science? Or will science subdue evolutionary theory? Science has no doubt benefited from Darwin's unsettling idea. Evolution as concept has taken over all of science. Star clusters evolve—so does just about everything. And evolution sparked remarkable efforts within biology. If Darwin hadn't considered discovering the mechanics of inheritance imperative, we might not have discovered DNA and decoded the genome. Evolutionary theory did what any theory is supposed to do: it energized investigation, sending researchers off in a multitude of directions and adventures.

Science still eyes evolution with a gimlet gaze. In fact, all of biology just doesn't fit the model quite the way it should. Darwin tried to make his theory fit, for he was under the thrall of Newton, as were all scientists of his day. Yet his work shows he knew he couldn't quite get it; you can detect a tone of regret in his gentle wordings. Instead he wrote *The Origin of Species* as one long argument, more like a lawyer's brief than a manifesto of immutable laws, revealing that his subconscious, at least, understood that organisms aren't objects and weren't likely to obey any laws he could write.

Darwin wrote a theory that advanced science and philosophy, but he was still a product of his own time. He probably did not know he was describing a world that was based less on certainty than on probability; he certainly wouldn't have described it that way. Other scientists were grappling with similar problems. Physicists, faced with explaining the behaviors of gases, began creating a science of the behavior of averages. Instead of trying to predict a gas by dissecting behavior down to the level of individual atoms (pictured as billiard balls bouncing off one another), Ludwig Boltzmann instead described populations as one entity. At the same time, mathematicians were working out the mechanics of probability and statistics. Quantum mechanics and Heisenberg's Uncertainty Principle were rewriting the science of

the twentieth century, making it into one of precise approximations. Maybe these approximations were the best that
one could get.

But even these new methods were too deterministic for
biology. The tiny units that make up a gas are still predictable, uniform, simple entities. Biologists like Ronald Fisher
tried to reduce their science to the most predictable units—
down to genes (not yet understood at the molecular level)—
but even then they buckled. Sewall Wright argued that organisms don't assort themselves uniformly, instead clumping
and associating in small, mobile populations he called demes.
And he insisted on a role for randomness.

To deal with the imprecisions, some scientists turned to
reductionism. They dissected biology down to its tiniest and
most obedient units—organic molecules. In the process
they created genetics and molecular biology. These fields
have yielded the successes of biotechnology and medicine,
as well as advances in basic science.

Scientists at home with the fuzzier logic of life remained
naturalists, at least to some extent. They stuck with "organismal" science, science starting with organisms and going to
populations and communities. For some, these less reductionist efforts seemed self-indulgent. The work seemed too
vague and anthropomorphic to be taken seriously. And it
wasn't likely to yield practical applications. Status and support dried up; funding was meager.

But recently, while molecular biology has not wavered
in its progress, its failings are becoming more apparent.
Even as advances continue to dazzle, it's possible the questions molecular biologists ask are too limiting, the solutions

they seek too narrow. They are busy enough, anyway. For the rest of us, it is clear that the living world supports too many phenomena that aren't explained by the diligent summing up of their component parts. An organism is more than a collection of genes, or even gene expressions. The result is something called emergence: when properties arise that exceed the sum of properties of the component parts. Life emerges; it is more than its mixed-up milieu of metabolic pathways and cell structures. Life comes about more as a result of organization itself, emerging from chaos.

Traditional methods of science, the steps we learned in school—observation, hypothesis, experimentation, conclusion—don't necessarily answer biology's needs. This classic model insists on constancy, unchanging parameters. That is not life. If we can hold a living system to such a static state, it is not a normal system. And our results cannot be normal.

So science, as we know it, has a problem with life. Some scientists are still trying to fit the round peg of life into the square hole of physics and chemistry (for that matter, both those sciences are challenging the system as well), but others are challenging that approach. We are left with a choice: Should we continue to try to lever biology into this old framework, or should the science redefine itself?

THERMODYNAMICS

As Huxley and Hooker were defending Darwin in the second half of the nineteenth century, physicists were elucidating the laws of thermodynamics. The first law states only that

157

energy is conserved and remains constant, but it changes and flows from state to state. This set up the need for the second law, which provided the "motive force" for energy flow. An early observer, Sadi Carnot, compared a steam engine to a mill wheel; the fall of heat from higher to lower temperatures drove the steam engine, just as the falling of the water turned the mill wheel. If the energy in the system remained constant, what explained the change in the system? Something had to be driving the change in energy from one system to another and physicists called that force *entropy*. In Carnot's mechanical systems, entropy was dissipated potential. If there is an imbalance between states of energy, something called a field is set up. The potential difference of that field generates entropy—the need to equalize and de-organize that potential difference. Natural processes occur because of the "need" to dissipate the potential difference of energy (and of matter, which Einstein showed us was just a different form of energy).

The second law of thermodynamics states that entropy is always positive, meaning that in a closed system it always increases. Stated most simply by Clausius in 1865, "The energy of the world remains constant. The entropy of the world strives to a maximum." Most of us understand it as the notion that systems tend toward disorganization, which is the interpretation Boltzmann used to explain the behavior of gases.

To describe it scientifically is harder, but worth trying. We know that when two solutions—one hot and one cold—are placed together, the molecules of each will mix, and the heat will dissipate. On a microscopic level the molecules spread out and mix randomly, becoming disorganized. Macroscopically, the random mixing makes the mix uniform—the whole solution comes to the same, lukewarm temperature.

This law of increasing entropy has implications for evolution. At first glance, life seems to contradict entropy because life is organized. Natural theology was all about how precisely well adapted living things are for their environments; this requires organization and the clever hand of a creator. Thus, that life could "just happen" without external intervention seemed impossible, and this seemed even more substantiated by the second law of thermodynamics.

On the other hand, it was increasingly clear that living organisms followed the law of entropy very well. Chemists were working out the steps of the respiration of plants and animals. They compared the process to combustion; all organisms did was burn the molecules available to them, slowly, step by step, and gather energy from this slow combustion. Metabolism was known to be a series of steps of molecular change, with each providing a dollop of energy, like Carnot's stream dropping into each cup of the mill wheel. If the organism loses these molecular cascades, the energy escapes and without it the complex organization crumbles. The system reaches maximal entropy—the organism dies.

So how could life occur in a universe where entropy ruled? How could organization come about without outside input? One answer to this came quickly in entropy's history from quantum physicist Erwin Schrödinger. In his book *What Is Life?*, Schrödinger pointed out that although entropy is irresistible for a closed system, it was possible to divert it, for it to spin off into what he called "eddies of order." That, he proposed, was life—from organisms to ecosystems; as long as the whole system was far from equilibrium (and the sun still providing energy made it so), these pools and eddies could hold entropy at bay, at least temporarily. What an organism feeds on, then, is negative entropy, what Schrö-dinger called "negentropy." He wrote, "the essential thing in metabolism is that the organism succeeds in freeing itself from all the entropy it cannot help producing while alive."

By not just subduing entropy but co-opting it, Schrö-dinger succeeded in keeping life within the framework of physical laws. In doing this he also started to free us from a static perspective. Instead of seeing the world as resting at some equilibrium point, he persuaded us to see that our world is far from equilibrium. The apple has not yet hit the ground; the pendulum is still swinging. Schrödinger showed that this is how life could exist in a world driven by entropy.

THE CODE

In his application for a research fellowship at Cambridge, Francis H. C. Crick wrote, "the particular field which excites my interest is the division between the living and the non-living, as typified by, say, proteins, viruses, bacteria and the structure of chromosomes." As history relates, he got the fellowship and experienced great success by applying him-self to just such a project. He admitted to having been inspired in part by Schrödinger, and by his own desire to dislodge religious claims. He wanted "to try to show that areas apparently too mysterious to be explained by physics and chemistry, could in fact be so explained."

Crick's target was deoxyribonucleic acid, DNA, known by then to be the medium of heredity. At Cambridge he joined a team with James D. Watson, an American encouraged to study genetics by Muller and X-ray crystallographers Maurice Wilkins and Rosalind Franklin. At first, the makeup of DNA made it seem too simple a molecule to encode inheritance, which many thought would more likely reside in the proteins also found in cell nuclei. But the computer age had started, and the simplicity of binary code systems made DNA a more attractive candidate.

Watson and Crick faced formidable competition in the race to elucidate the structure of DNA. Linus Pauling, a California Institute of Technology biochemist who had already won a Nobel Prize for work on cell respiration, was reputed to be working on the problem. Pauling revolutionized the chemistry of life by applying quantum mechanics. He drew molecular reactions down to their electron orbital interactions. Faced with such competition it seemed two mere graduate students didn't stand a chance.

Watson described a meeting with Pauling's son in England. He carried a manuscript rolled up in his jacket pocket they suspected was his father's latest attempt at DNA structure. There they were, all in one room, Watson and Crick certain that the manuscript spelled out their own doom, dying to see it while pretending to maintain scientific objectivity. They did finally get a look, and it turned out that the model described in the paper was a triple helix. Watson and Crick, who had already tried this model and discarded it, breathed tremendous sighs of relief. They weren't beaten after all.

The two had the tremendous benefit of working with Rosalind Franklin, a meticulous and brilliant X-ray crystallographer. Her work was impeccable, but she lacked Watson's and Crick's audacity. They were willing to extrapolate findings into model after model when she preferred to wait for more evidence.

Their strategy worked, if only by a whisker. Their paper announcing the double helix, submitted October 14, 1952, was published in the November issue of *Nature*, while Pauling's, submitted October 22, appeared three months later— in January. The two papers were submitted at nearly the same time, but *Nature* tended to publish papers from England (being a British publication) as soon as possible. Pauling's paper also required the production of diagrams, which took some time to prepare.

The importance of this structure was not exaggerated, for structure was the immediate clue to function. There were four bases of DNA, which could be inserted completely interchangeably anywhere in the single chain molecule. The closed ladderlike structure allowed the molecule to be sealed and then reopened as needed. It was a lovely, neat system, clearly perfect for the duplication and coding of information. Watson and Crick introduced it as having "novel features which are of considerable biological interest."

Little time has been lost since then. The code has been deciphered. Different three-base combinations stand for various amino acids; the amino acids are linked together to form the functional proteins of cells and organisms. The mechanism of reading the code and implementing its instructions has been worked out, as well as the mechanics of copying it. And now we have read the book of life, as it were, and have catalogued the instructions for building humans, yeast, and fruit flies.

In the second half of the twentieth century, biology was nearly overwhelmed by the findings of genetics and molecular biology. Advancements came so quickly, and were so important, it seemed as if there was no other part of biology worth pursuing. Life really could be dissected down to its molecules, and doing that was the way to parse its secrets. Genes, hormones, neurotransmitters, cell mediators, antibodies—all were chemical units that drove so much of the machinery of life. It really seemed that it would be only a matter of time, and tedious sequencing, before life would unfold before us like a flower blooming. As a consequence, biology lost sight of the whole organism. Whole organisms were too big, too clumsy, and too hard to predict to be important study subjects. It took a while before science started to realize it was missing the forest for the trees.

GAMES

While funding went into the pockets of geneticists and molecular biologists, a few crusty naturalists stubbornly clung to their own preferences. They didn't roll over and just accept their own obsolescence. Instead they brought in their own kind of modernity. Some trained in the now-old tradition of biometrics now used a new kind of math: game theory.

The old arguments against the role of natural selection were revitalized by Motoo Kimura's neutrality theory. With mutations now seen to be mostly neutral in effect, chance distributions of mutations and genetic drift became more important drivers in evolutionary progress. Selection and adaptation started to look less important. In particular, for more traditional Darwinians, the inability to explain how some behaviors evolved haunted them.

With altruism, "selectionists" faced one of their biggest challenges. In humans we think of heroism and self-sacrifice as altruistic behavior; in animals altruism is less self-conscious but no less self-sacrificing. Altruistic acts by animals include maternal care and protectiveness, and alarm calls and displays. The loud cry of a bird or the great leaps of alerted impala attract the attention of the predator, making the alerted individual a more obvious target. Even the colonial life of social insects is considered altruistic, for the workers sacrifice their own reproductive potential in order to raise the offspring of their queen. All of these behaviors are hard—if not impossible—to explain if you use models emphasizing competition between individuals.

Redirecting selection's impact from individuals to groups made up one early attempt to explain the evolution of altruism. The Scottish ecologist E. C. Wynne-Edwards posited "group selection" as an explanation for altruism. As an ecologist, his perspective was already community oriented. He proposed that if the fitness of the group was improved by the sacrifice of a few individuals, the result would be longer survival for the species. He used this premise to explain the observation that some populations appeared to limit their own populations when resources were scarce by choosing behaviors that reduced some individual's reproductive rates.

The idea was appealing, but the models did not hold up to scrutiny. One biologist in particular, George C. Williams, was particularly scathing. Cheaters, he pointed out, would always prosper (and survive and leave more offspring), by the actions of altruistic individuals. The altruistic members of a group would be quickly eliminated and replaced by the cheaters. Evolution would always favor cheaters and altruism would disappear.

But not, as biologist W. D. Hamilton showed, if the altruistic members were related to one another. He studied the social insects (Hymenoptera), tracing their genetic histories. He discovered that the workers, who were always female, were more closely related to each other than they were to their own parents or male siblings. Sisters invariably shared more than half of the same genes. Thus, if they wanted to immortalize their own genomes, it paid to help their mother produce more sisters. Hamilton's "kin selection" provided a dramatic solution to altruism.

The same sort of strategy, using genetic relatedness as the key to fitness and reproductive success, could be extrapolated to other species. Although other species might not share the same unequal distribution of genes as did Hymenoptera, there was still some calculable benefit to promoting the survival of kin, who no doubt shared some of the same genes. J. B. S. Haldane once commented that he would willingly die for two brothers or eight cousins—what he was doing was calculating the odds that those particular combinations of relatives would likely contain most, if not all, of Haldane's own genes.

Kin selection has proved to be a particularly productive idea. Merging kin selection with the mathematics of game theory accounted for a multitude of behaviors that couldn't be accounted for with classic selection models. In recent decades the advent of DNA matching—where two organisms' DNA could be compared and relationships analyzed—has led to even greater understanding of the role kinship plays in behavior and evolution. Being able to measure relatedness has shown that in many species behavior is calibrated accurately to kinship: alarm signals, cooperation all correlate closely according to the relatedness between individuals in many species.

More Games

Another example of game theory applied to evolutionary theory comes from John Maynard Smith. Smith originated the term Evolutionary Stable Strategy, also known as ESS. An ESS is just what it sounds like, a set of characteristics, physical or behavioral, that support and maintain the reproductive success of a species. The following is an example of working out what an ESS might look like:

Say a single species includes individuals with opposite competitive strategies. Some individuals' strategy is to be a "hawk." A hawk if challenged will fight even to the point of serious injury, even when that injury resulted in a decrease of fitness. The second strategy would be to be a "dove." A dove will retreat in the face of challenge always, to the point of reducing its fitness because it cannot hold claim to resources.

Smith then built a checkerboard (more properly called a matrix) that shows all the possible outcomes of interaction between these individuals. What the matrix reveals is that if all the individuals are hawks, the species would increase its risk of going to extinction. This would happen because, as so many individuals became injured, too many would have decreased fitness, resulting in a decreased effective (reproductive) population. As the effective population falls, the species becomes more and vulnerable to being wiped out by a random event—an epidemic, fire, storm, etc. If all individuals were doves, the same decrease in fitness and in effective population would occur. A population of all doves would also be likely to become extinct.

The most evolutionarily stable strategy for the species would be to have some of each. A sprinkling of hawks among the doves or vice versa showed the most promise for the most offspring for the species as a whole.

Unlike kin and gene selection theory, this is selection on the group or species level. The characteristics described by Smith would impart advantage for the group or species as a whole and not just for the individuals—the individuals themselves would also be advantaged. (Also, the hawks would always win against the doves, but the doves would survive as long as the hawks had hawks to compete with and do damage to one another.)

SOCIOBIOLOGY AND GENE SELECTION

Kin selection spawned a synthesis of evolutionary theory and behavior called sociobiology. Outlined in a book by Harvard ant biologist Edward O. Wilson called *Sociobiology: The New Synthesis*, it promised a new way to understand evolution. This was a Darwinian theory of evolution—adaptation and selection were paramount forces. The idea of kin selection has evolved to be an explanation for more globally observed phenomena. For example, it has been used to explain the evolution of the intelligence of primates. Monkeys, apes, and man needed the extra brainpower, it is suggested, just to keep track of who they were and weren't related to, and by just how much.

A second, smaller, but just as influential book took kin selection one logical step farther. In the tradition of reductionism, Richard Dawkins took the obvious and elegant step and wrote *The Selfish Gene*. In this he makes the case that what nature selects is not the species, the group, or even the individual but the gene. A gene for altruism, to put it in greatly oversimplified language, is reproductively fit, because one of Haldane's two brothers or eight cousins is sure (statistically) to carry it.

Dawkins's system is elegant and simple and can be reduced to a sort of law of self-perpetuation. Things, not just organisms but genes—simple segments of DNA—survive and last because they are stable and are replicated. They are replicated more when they have characteristics that make their organism more successful than other organisms. In this system the organism becomes just a "survival machine" for its genes. The survival machine doesn't last, but the gene does. In Dawkins's words: "Genes, like diamonds, are forever. Individuals and groups are like clouds in the sky or dust storms in the desert."

This scenario explains a puzzling phenomenon of genomes. Within a genome, there is often a great quantity of (allegedly) useless DNA. These are stretches of DNA that don't seem to code for anything. Dawkins's selfish gene theory could then be reduced to selfish DNA theory. This DNA survives simply because it can. It becomes a "parasite" hitching a ride on some other DNA's survival machine. That the same useless stretch of DNA might later become a useful gene, after being slowly altered by mutation, just fits in as a special case.

It is important to recognize that the word *gene* takes on a new meaning in Dawkins's context. It no longer means just a length of DNA, nor does it mean a code for a specific protein. Dawkins uses the term to describe something more nebulous and complex. He is using gene as shorthand for a trait or characteristic, often a behavior. He is quick to point out that a single gene in the molecular sense is probably not the progenitor of these traits. Instead, some unknown quantity of genes or set of gene interactions form his gene. He is often accused of oversimplification and ignorance of how genes actually work. This is unfair. He doesn't claim to be talking about genes in their molecular incarnation—he is using the word gene in the abstract. In a way, he is not insisting on genes as the selective unit, but does insist genes are the originators of the selective unit.

If Dawkins's genes are abstractions, his next extrapolation is even more so. In the same spirit, Dawkins proposed the existence of "memes." Memes are more or less cultural units of selection, social entities that exist because they have survival value. Societal rules, organized religion, public education exist because, again, they are self-perpetuating and the organisms that use them experience some increased survival value as a consequence. According to Dawkins, memes bring evolution to a new level, sort of meta-evolution.

BACKLASH

With gene selection and sociobiology, some felt evolution science had gone too far. The informed general public and some scientists reacted with a very fervent and public backlash. At the peak of the controversy, as E. O. Wilson rose to speak at a meeting of the American Association for the Advancement of Science, protestors ran up and poured a pitcher of ice water over his head.

The protestors were mainly reacting to the last chapter of Wilson's *Sociobiology*, in which he extended his synthesis to include the social behavior of humans. He dared to suggest that even humans could trace their intriguing social characteristics to an evolutionary source. For some this was too reminiscent of social Darwinism and eugenics. It was the old argument of nature versus nurture, and Wilson's claims sounded too much like genetic determinism. It was particularly offensive to feminists who believed the theory could be used to substantiate claims that women were not biologically suited to certain careers.

Some of the protest was a battle over academic turf. Sociologists and psychologists saw sociobiology as saying that upstart biology had more to contribute to the understanding of the human condition than they did. (And even though biology was seen as a "soft" science, it was much "harder" than the social sciences, especially since it could lay claim to molecular biology. So from the point of view of the psychologists and social scientists, sociobiology could be interpreted as just another land-grab by the scientific establishment.)

Academic objections were not all interfactional; evolutionary theorists of a certain ilk took umbrage as well. The loudest critics tended to be social leftists, who found fault with sociobiology because it seemed simply capitalist in character. One in particular, Richard Lewontin, was a Marxist. He was the experimentalist who, as Dobzhansky's student, used protein electrophoresis to demonstrate the extreme genetic heterozygosity of fruit flies. Like Dobzhansky, Lewontin was strongly opposed to eugenics and rejected Fisher's gene-based models for the less rigid, more chance-driven models of Wright's.

But Lewontin did not reject Wilson's sociobiology and Dawkins's gene selection merely on political principle. He had studied and worked with Wilson, and knew Wilson was not promoting eugenics and genetic determinism. He still insisted the trend was dangerous, and refused to separate the theory from social consequence. He believed that the directions the theories took, the justifications for them, would taint thought and move it in the wrong direction.

As a scientist, he knew better than to base his criticisms on these premises. He pointed out that in development, organisms are strongly affected by environment and are not the perfect embodiment of their genes. He gave the example of the eyes of fruit flies. Normal fruit fly eyes are constructed of about 1,100 facets, and this does not change if the eggs are incubated at 15 degrees Centigrade. Incubate the same eggs at 30 degrees Centigrade and the eyes will develop with only 750 facets. He demonstrated the same variability in mutants that developed eyes with fewer facets. The same alterations in incubation temperature would elicit similar differences in facet production, although the mutants would always have fewer facets than the normal flies. It is impossible, Lewontin pointed out, to predict what an organism will look like if all you know about it is its genome.*

Somewhat surprisingly, Lewontin found an ally in Ernst Mayr. Although far from a leftist, Mayr found gene selection a bit hard to swallow. Genes can never be the focus of selection, he argued, because the genes are carried by the

---

* In a similar fashion, cloned organisms are not identical replicas of their "parent." The black-and-white patterns of Holstein cattle clones and the coat colors of the first cloned cat are quite different from the originals.

organism. The characteristics of the organism develop from thousands of genes and probably millions of gene interactions. Thus the gene is never acting alone and so selection can never act on a gene alone. Nature does not impact a single gene—it impacts the collective actions and interactions of the organism's entire genome. No matter how fit one gene may be, if the organism it is in is not fit the gene will be lost.

(Mayr may have also objected to what he guessed was an attempt to more tightly shoehorn biology into traditional science. Mayr has been quite energetic in his arguments that biology is its own discipline and correct to disregard the more formulaic approaches of chemistry and physics.)

Lewontin went on to make a strong case for the existence of group selection. He used house mice as an example. House mice tend to live in small populations, called demes, which are reproductively isolated from other demes. House mice also carry a gene, called the t allele, which is lethal in male mice carrying two such alleles (being homozygous for the t allele). In a deme, male mice who carried a single t allele and were otherwise reproductively successful, could produce enough t alleles to result in a great many t allele homozygous male mice offspring, who would then die. This scenario could effectively doom the deme to extinction. (And doom the t allele to extinction, as well.)

But the t allele survives in spite of this, because at the same time, it uses a trick to promote its own survival. Within the germ cells, the t allele makes copies of itself and inserts extra copies into the genome, in effect creating more heterozygotes that happen to carry the t allele, preserving it. As a result, the survival of house mice depends more on the characteristics of the genomes within each deme; the success of mice carrying the t allele must be counterbalanced by the death of the homozygous males, which must be in balance with the rate the t allele copies itself. A male mouse in one deme, then, can be more evolutionarily fit than if it were in another deme. Examples of similar genetic idiosyncrasies in other species have supported the strength of this argument.

PUNC EEK

Another group weighed in against gene selection, this time from a usually more modest quarter. If molecular biology had muffled the voice of organismal biology, it had completely silenced paleontology. Scientists who studied fossils were thought of as little more than parlor entertainers and stamp collectors, fit only to be museum curators. Most paleontologists quietly accepted this role. If they observed anything in their fossils that didn't fit the model of evolution handed down to them by geneticists and molecular biologists, they kept it to themselves.

They did, that is, until Stephen Jay Gould and Niles Eldredge came along. These two young fossil hounds thought they would rock the evolutionary world by airing a rather dirty little secret. The prevailing opinion in the 1970s was that evolution was indeed, as Darwin described it, gradual. The discoveries of genetics and the New Synthesis supported this conclusion. Gould and Eldredge had different evidence, and they weren't about to let it stay overlooked.

They knew what other paleontologists knew: that fossils from one site looked the same for millions of years. From this it appeared the species remained constant; forms and sizes of organisms were static for geologically long periods of time. This in spite of evidence showing that the environment changed, even to the extreme conditions of ice ages. Then, suddenly (in geologic terms), radically different fossils would appear, different enough to be completely new species. Gould and Eldredge called the alternating of stasis and rapid change "punctuated equilibrium," nicknamed Punc Eek.

Evolutionary change in such a mode did not fit the New Synthesis, but instead suggested change might be more rapid, more like the saltation supported by Bateson. With this model they were challenging so called neo-Darwinism, although both Gould and Eldredge considered themselves Darwinians. Neither Gould nor Eldredge proposed new models for the act of creating new species. All they were doing is pointing out that the historical evidence looked stronger for rapid evolutionary change.

Reaction to punctuated equilibrium as a model for evolution has been tepid. Most evolutionary biologists didn't react at all at first, and wouldn't have given it much thought except for one thing. Gould was a remarkable and prolific writer and an effective popularizer of science. Even if the scientists hadn't read his work, they couldn't help but hear about it.

So they have tried to explain it away. Some have cited the patchy nature of fossil formation, the paucity of transitional forms—just as Darwin did. They cited a few cases where, just by traveling a short distance from the original site of a given fossil, all kinds of intermediate forms could be uncovered. The problem was, not enough of those second sites have been found.

Time itself forms one of the strongest arguments against the significance of punctuated equilibrium. When fossils do change in the fossil record, although it seems to happen in a *geologically* short time, that does not mean the period is *biologically* short. It still may be enough time for the gradual change predicted from gene mutation and recombination.

So goes Mayr's argument. Mayr is best known for his theory of "allopatric speciation" (*allopatric* means "another country"). Mayr theorized that the evolution of new species will most likely occur when a small population becomes geographically separated from its parent population. With a small number of migrants, there would be a smaller selection of genes for the new population, allowing it to veer off, so to speak, from the original population. Mayr called this the Founder Effect, for the limited numbers of founding members of the population. If the group was also subjected to intense selection pressures, for example, a new environment or new competitors, the change could be rapid— rapid enough to fit into Gould's scheme of punctuated equilibrium.*

---

* Theoretically, according to Mayr, sympatric or "same country" speciation could also work. It could happen if a population became marginalized, or if some members became reproductively isolated. For example, a female peacock could decide she only liked males with purple eye spots on their tail feathers. She could "found" a sub-population within the geographic range of the larger population. Mayr could model this event, but noted that there was no evidence that it had ever actually happened.

With allopatric speciation, the new population could succeed and multiply. Since it perhaps evolved some advantage over the original population, it could migrate back to the original range and out-compete, and possibly replace, its parent population. If, instead, the population returned before the two populations changed enough to create reproductive isolation, the two populations would merge. The resulting population would absorb whatever genetic advantages the migrant population returned with. The species would have evolved, but would not have become two species. Or, according to Mayr, one must take into account the dimension of time. The changed population may indeed have become a new species, one that would not be able to merge with the original species if the two existed at the same time.

This scenario explains most of the observed fossil record, but there is a logistical problem. It becomes hard to explain why species experience evolutionary stasis (equilibrium) for millions of years, especially when the environment is changing at the same time. Evolutionists have proposed that organisms experience genetic corrections, called "stabilizing selection," where any new versions of genes and phenotypes are quickly eliminated because they are non-adaptive. This explanation fits with Dobzhansky's theme of balancing selection, where genotypes are so interconnected and related that they resemble a sort of Chinese finger trap; any change results in a warping of the entire organism.

There are problems with stabilizing selection. It presumes a frenetic but underground activity going on constantly in a species' genetic makeup. Leigh Van Valen called it the Red Queen hypothesis, after the Red Queen in *Alice in Wonderland.* In that book, the Red Queen and Alice are running together, the queen urging Alice to run faster and faster. But Alice can see that the scenery is not changing— they're not getting anywhere. When she points this out to the queen, the queen responds, "Of course not; we have to run and run this way just to stay in one place." Many biologists find this scheme hard to accept. If species experience such frenetic turnover of their genomes, why aren't there more noticeable outward changes? And if the environment is also changing, it makes the species' lack of change all the more remarkable.

Gould calls up history to offer alternative theories. In the 1940s a developmental biologist, Richard Goldschmidt, proposed that the creation of new species, macroevolution, must require a mechanism different from microevolution, change that occurs within a species without creating a new one. He theorized the existence of "rate genes." These rate genes would be turned on or off by some signal coming from the environment. The effect would be to alter timing, or rate, of development, and would result in the dramatic changes to an organism's phenotype. Thus would be created what Goldschmidt called "hopeful monsters"—trial organisms testing new fitness strategies against the altered environment. Most would die, but perhaps some would succeed and live and become new species in one fell swoop, more or less.*

---

* These suggestions echoed some of Darwin's ideas. Darwin made several efforts to understand speciation, finally favoring one that postulated that environmental disturbances during development created new species.

His work was condemned and disregarded, because some of the details he'd suggested turned out to be wrong, but also because he wrote during the time of the New Synthesis. That was a time when developmental biologists were supposed to sit back and wait for geneticists to make the real progress. His ideas were resurrected briefly by Conrad Hal Waddington in the 1950s. Waddington did experiments to demonstrate that environment could indeed have profound effects on embryonic development. He subjected fruit fly eggs to thermal shock, inducing mutant strains. The changes were permanent and he called them "genetic assimilation." He extrapolated his findings to draw new models of how genes influenced development, emphasizing intricate interactions. He called these models "epigenetic landscapes," after Wright's evolutionary landscapes. Perhaps, he said, each organism contained the genetic instructions for more than one phenotype. Perhaps something that happens early in development results in a switching, like a train from one track to another, from one phenotype to another.

ADAPTIONISM

Goldschmidt and Waddington, followed by Gould and Lewontin, were suggesting that organisms themselves play a role in determining the direction of evolution. Selection does not act upon the organism independently of the organism's basic design. Nor does it act directly on the genes as independent entities. Genes, they were saying, are not omnipotent. They insisted that some of the influence must come from the organization of development itself.

This meant throwing out the rigorously adaptionist nature of gene selection. Adaptionism itself they found objectionable. The tendency of evolutionary theorists to explain every characteristic as an adaptation was, they said, merely knee-jerk reductionism. They compared it to the *Just So Stories* of Kipling or the feel-good justifications of Dr. Pangloss in Voltaire's *Candide,* and to Paley's natural theology. Not everything, they said, was created to perfectly answer some need. Instead, organisms looked the way they did simply because they did, because they fit together that way, because they didn't have the means to look any other way. As Gould put it, nature is a tinkerer and works with the materials at hand. Luck plays a role as well. Perfectly fit organisms could still be wiped out by cataclysms. Less fit organisms could survive by being in the right place at the right time.

One metaphor for this, from Gould and Lewontin, is famously known as spandrels. Spandrels are concave triangular panels created in cathedrals at the intersections of support arches. Within the spandrels artists created elaborate works of religious art. One might look at these works, the authors say, and guess that the spandrels were made so that the art could be put there. Not so. Spandrels just happened because of the need for arches. Only the subsequent decorations made them look planned (or adapted).

Likewise, they go on, there are no genes for chins. Instead a chin forms because two halves of a mandible come together there. A chin can be large, small, dimpled—none of these variations is adapted for anything. The chin just happened.

DESIGN BY PROXY

This idea of constrained development can be taken further. Perhaps there are patterns in nature that impose constraints on form. Physical size is one such constraint: elephants can't produce wings large enough to allow them to fly, nor can their legs be thin and spindly. The evolutionary commitment to a body plan imposes limits; mammals have never developed an extra sets of limbs, for example. The suggestion is that at some point in evolution, and in development, a fundamental body plan is locked in. Evolution can modify the plan, but not start over with a new one. For that matter, a species cannot recreate itself—once extinct, it doesn't come back.

Brian Goodwin has taken that idea into the realm of complex chemistry and mathematics. He has demonstrated that forms in nature reveal the existence of fundamental patterns. For example, the distribution of leaves and flower petals in plants fit the predictions of a thirteenth-century mathematician, Leonardo Fibonacci. Fibonacci wrote a series of equations intended to predict the fecundity of rabbits. Since then his equations have been rediscovered, and have been seen to apply to many other biological phenomena as well. As Goodwin points out, Fibonacci equations generate series of numbers that correspond to the intervals and angles of the positions of leaf buds and flower petals on

181

plants. Equally fortuitously, the patterns of slime mold growth closely resemble patterns generated when chemical reactions are imaged in progress as they occur, in work done by Russian chemists Lev Beloussov and Anatol Zhabotinsky.

These serendipitous resemblances may mean that nature imposes shape and patterns on organisms. If this is true, it leaves genes with much less responsibility. Ditto natural selection. As Goodwin writes in *How the Leopard Changed Its Spots,* "Natural selection does not have a lot to do except act as a coarse filter that rejects the utter failures."

TELLING STORIES

Punctuated equilibrium also gave credence to another concept promoted by Gould, that of a hierarchy of selection. Rather than trying to find some fundamental unit for selection to act upon, which is perhaps too simplistic a model, why not simply admit things aren't that simple? Why not admit that selection can act on genes, individuals, groups, kinships, and indeed even species? Gould, recalling Wallace, included species among the entities selection acted upon.

Ecologists carry the hierarchy out even further. To understand whole ecological systems, ecologists see evolution affecting all of it. For them, whole communities respond to environmental change or to smaller changes within the system. A faster impala leads to a faster cheetah, or a cheetah that eats more wart hogs and wildebeests, or both. The effects of any change are not isolated, but ripple out, touching many communities. And you can't separate a single ripple; you can't even separate the pond from the shore—each particle influences every other.

The principle of hierarchy of selection added to evolution another tier of complexity, far away from the reductionist influence of Dawkins and far away from the framework of the more physical sciences. If selection works at all these levels, and combines with system constraints and chance, then evolution will never be explainable in terms of simple, rigorous laws and principles. Gould, Mayr, and ecologists assert that biology is not given to the same framework as physics and chemistry. As a science, it is too complex for that.

They suggest a better way to study biology may be through narrative. Scientists should not expect to be able to explain biology in terms of laws and formulas. Perhaps the way Darwin presented his ideas, using description and argument, was after all the most appropriate way to approach it.

Ecologists have opted for this solution for practical purposes. Faced with the job of assessing human impact on the natural world, they can't predict a single scenario. If they try, they'll be wrong too often and no one will listen to them. Complexity going in produces complexity coming out; if ecologists build their models accurately, the multitude of modifying elements produce a multitude of possible endpoints. Since their models can come up with so many different scenarios (depending on what the inputs are and what they can make of the relationships between the inputs), they have instead adopted the use of "if this, then that" narratives, or worst case and best case scenarios. Asked to contribute to policy decisions, the advice they can offer is

183

in the form of stories or descriptions. They no longer can allow themselves to be pinned down to predictions. Far better to keep future plans flexible and contingent—as things go awry, they have a plan ready and waiting to deal with possible consequences.

To suggest that science is telling stories—well-grounded and well-formulated stories, at least—is not the laugher it might have been. The ecologists call this post-normal science.

LIFE AS DISSIPATIVE STRUCTURES

The second law of thermodynamics takes science into the realm of the nonlinear. The trajectory of a falling apple, or the arc of a thrown ball are constant, predicable, *linear* events. They are predictable precisely because there are a limited number of measurable inputs: the strength and direction of the throw and the force of gravity.

The chaotic, the complex, the catastrophic are all nonlinear. They include systems like weather, business cycles, the heart, and the brain. These systems have vast sets of inputs, many of which cannot be accurately known. One can only model these systems using computers and a new kind of math. They require a new set of expectations. The mathematics predicting how all these systems will behave produce not pinpoint answers, but a range, something between zero predictability (completely chaotic) and dynamic systems of probability. The dynamic systems, when mapped out by computers, look like three-dimensional blobs. But they have some predictive power—they tend to orient around a distinct region or point. Those regions or points are called attractors.

For example, a pendulum will swing in imperfect circles, but will tend to fall toward a point of rest, called a point attractor.* If you design a more complicated system with more inputs, the computer generates a three-dimensional picture that looks something like a kidney bean, with the system tending toward two specific regions. The graph will look like a tangle of strings, but the strings will circle or swirl around two more or less circular regions. Real life looks this way: business cycles look rough and jagged and unpredictable, but they tend to repeat themselves. The weather changes and we are caught without an umbrella, but less often in some seasons than others. We know that change happens, but in a vaguely regular fashion.

Back to thermodynamics. All this gets back to fields or gradients—one end of the system has a lot of something (heat, entropy) and the other end has only a little. The abundant, as the second law tells us, tends to diffuse toward the nonabundant, the system strains toward an equilibrium state of microscopic mixing and macroscopic uniformity.

That sorting out can be completely chaotic, but often it is not. Instead it is complex, vaguely predictable. For example, put a pot of water on a hot stove. As the heat enters the system, it creates a gradient, or field. Water molecules at

---

* Another example of a self-organized structure is the stormy red spot on Jupiter. It perfectly matches Earth-bound models for dynamic gaseous fields organized around a point attractor.

the bottom, close to the heat, become active and rise, more or less willy-nilly, to the surface, where there is no heat. As the heat increases, the gradient becomes greater. But this is a complex system—water molecules have to deal with the heat *and* forces of mutual attraction *and* viscosity. With the added inputs comes added constraints. With all of these influences, water molecules simply can't rise fast enough if they remain completely unorganized.

As a result, convection currents form. As the water begins to boil, the bubbles come up in a regular honeycomb pattern. The currents look like elongated donuts, the water molecules rising and falling back down following essentially circular paths. This semi-organized behavior allows for more efficient dissipation of heat energy.

These sorts of self-organizing, dissipative structures occur all over in nature: hurricanes and tornadoes and the vortex of water running down your tub drain are all dissipative structures. Maybe living beings are just elaborate entropy-dissipating structures.

The elucidation of these structures induced Ilya Prigogine, a Russian-born physical chemist, to assert that the second law of thermodynamics is not just consistent with evolution, it helps explain it.

POST-POST-NORMAL

In his second year of medical school at the University of California, San Francisco, Stuart Kauffman was bored. He was interested in embryology, but mainly because he wanted to work out how natural selection might act on developmental steps he was studying. He had read theories of how regulatory genes might function in parallel rather than sequentially, and he was intrigued with the idea.

Kauffman had a background in philosophy and understood systems of logic. He decided to try to fit gene function into a Boolean network. Boolean networks harken back to George Boole, an English mathematician of the nineteenth century. In Boolean networks, elements can be in one of two states: on or off. Which state they are in depends on the activities of various numbers of modifiers. For example, an element may be on if one modifier is on, or it may be on only if *all* modifiers are also on. There are sets of rules that determine these contingencies, and all can be programmed into simple computer systems. For Kauffman, the elements were genes, which were either expressed or not expressed. He set up his systems to have varying numbers of genes and modifiers for the genes.

Once programmed, all the permutations are tested and graphed. Depending on numbers of modifiers, Kauffman's models yielded two distinct, if complex, sets of results.

First, with very few modifiers, the results form a three-dimensional graph with lines circling around a few regions. These regions are the attractors; the results tend to be close, or attracted, to the results represented by the region. If he disturbed the system, the graphs would change but then settle back into cycling around the same attractors. From this Kauffman could see that genes subjected to few modifiers would tend to produce organisms with a predetermined set of characteristics. This model suggested that constraints on evolution were built into the system—that evolution was determined more by the original form of the organism than by any external forces. This looked bad for selection as an important force in evolution.

Kauffman took the model a step farther. He added more modifiers, just enough to bring the model to the edge of chaos. If he pushed it all the way to chaotic activity, there would be no predictability at all, only chaos. But Kauffman stopped at the edge, and there he found a very different picture.

There were attractors at the edge, but they were not consistent. That is, if he disturbed the more complex system in any way, it would settle down again, around the same *sort* of region as in the first model. But the attracting region would not be in the same place at all. A disturbance in the system could result in a new place of equilibrium—a new form—and it was quite a different place from the original.

This meant that external change could have a strong effect on the system. Or, in a model of genes and organisms, that the environment could drive significant change as long as the system was on the edge of chaos. As he manipulated this model more and more, he found that it liked being on the edge. Indeed, Kauffman concluded that selection itself, outside disturbance, tended to keep the system on that edge, so selection essentially selected for itself. In this way Kauffman, who at first gathered evidence against the Darwinian theory of evolution, in the end found himself with a system much more compatible with Darwin's ideas.

If all this seems too abstract and rather far from reality, you are not alone. Kauffman and others are squeezing themselves into a new niche, making a place for completely theoretical biology, analogous to cosmology's place in physics. John Maynard Smith even goes so far as to disparage it as a "fact-free science," since it exists only within a virtual realm.

Still, it offers a provocative new paradigm that seems smoothly continuous with the past: Darwin started biology on its path away from the static predictability of Newton; early population geneticists carried it into the realm of probability, as quantum mechanics carried physics into a world of uncertainty. Darwin took biology away from the static mechanics of Newton; thermodynamics moved it far from equilibrium; complexity carries it even farther, to the very edge of chaos. There is no longer one answer, not even a range of close guesses. Instead there are constraints and propensities, and that is what we should expect; that is the way we should now perceive the world. Life is neither predictable nor regular—not perfectly random, nor perfectly designed. As David Depew and Bruce Weber put it in their book *Darwinism Evolving*, "we now recognize that, in spite of what Einstein believed, God not only plays with dice, but the dice are loaded."

Glossary

adaptation   The process that leads to the optimization of every part of an organism. Adaptationism can be compared to the philosophy of Dr. Pangloss, a character in Voltaire's *Candide,* who believed that everything (including bad things) happened for a purpose in this "best of all possible worlds." This view was criticized because it could lead to such statements as the reason we have noses is to keep up our eyeglasses.

agnostic   The word coined by Thomas Huxley in 1869 to describe a person who asserts that it is impossible to know if an ultimate cause (e.g., God) exists or not.

allele   A kind of gene. Specifically, an allele is an alternative form of a given gene, making it slightly different in form and possibly function. The gene for eye color, for example, will have two alleles, one coding for brown eyes, the other for blue.

allopatric speciation   The theory by Ernst Mayr describing the evolution of a new species as occurring after the geographic isolation of a small population from its population of origin. Part of the speciation would come about because of founder effects: the establishment of a population by small numbers of individuals with limited amount of genetic diversity. Selection would act upon the population more profoundly, perhaps, both because of its limited diversity and because it is subject to a new environment.

animalcules   Microscopic organisms, particularly motile protozoa and protists such as paramecia and rotifers. Word coined by Anton van Leeuwenhoek, who in the 1700s was one of the first to explore this tiny world with a microscope.

archaeopteryx   A fossil of prehistoric bird having characteristics similar to reptiles, including a pelvic bone and tail formed like a reptilian pelvis and tail, claws on each wing, and teeth. Found in Germany in 1863 and claimed as a transitional form by Darwin's supporters.

atheism   The belief that there is no god or gods.

attractors   In complex and chaotic nonlinear systems, a preferred region or regions within which predicted results tend to fall. The best example is to imagine a pendulum suspended from a single point on the ceiling and set in motion. Lines drawn by the pendulum will center, circling around a point where it will eventually come to rest. That point would be called an attractor.

biometricians   biologists who emphasize the use of mathematics as integral to the study of life.

Boolean networks   George Boole in 1854 combined algebra and logic to devise a system for describing systems with numbers of elements and inputs, in which the variables of interest change in binary fashion (on or off). Used by Stuart Kaufmann to model how regulatory genes can affect structural genes. Example: Kaufmann could posit that a structural gene would be turned on if one, some, or all regulatory genes provided input.

catastrophism   The theory of the formation of the physical earth based on large, cataclysmic events, such as a worldwide flood. While catastrophism concurred with biblical writings, the theory was put increasingly in doubt by the construction of canals in Britain, which, by revealing stratification of deposited layers of sediment, demonstrated the gradual and prolonged changing of the earth.

chaos   Unpredictability. A chaotic system is one in which any external change can cause unproportional (thus unpredictable) and significant change in the system. On the other hand, such changes will fall within a finite range, so that it is possible to predict the effects statistically. Life and evolution of life are good examples of chaotic systems, where there are tremendous numbers of inputs and often the effects of those inputs do not fit into simple, linear analytical systems.

chromosomes   The first microscopically visible manifestations of heredity, chromosomes are a combination of DNA and proteins that resemble bloblike Xs.

common origin   The claim that all life originated from one event, or one set of events, that happened in the distant past.

complexity   Any system responsive to the interaction of many parts, including circular or feedback interactions, so that linear, directly proportional predictions become difficult.

Also refers to self-organizing structures that occur in chaotic systems.

creationism   The belief that life and species were created by an agent. Contemporary creationism varies from the claim that the universe was created literally as explained in Genesis to the belief that the biblical description is metaphorical, but that God did create the earth and all life.

deism   The belief that God created special conditions that resulted in the development of the world and nature. In other words, God built the watch, wound it up, and let it go. (Or God is the absentee landlord.) This belief would allow a role for free will.

demes   Population units, also called effective populations, meaning only reproductive individuals.

deoxyribonucleic acid (DNA)   The molecule formed of four bases organized in a regular linear molecule that binds in a mirrorlike fashion to another DNA molecule. The two form a sort of twisting ladder, called a double helix. A single DNA molecule codes for the production of proteins by the order of its four bases—each three bases coding for one particular protein component called an amino acid. The result is that DNA is a sort of blueprint issuing instructions for the growth of an organism. Each organism's DNA is unique, so that each organism's makeup of proteins and structure is unique; thus DNA is the molecular basis of inheritance.

developmental biology   The study of the stages of development of organisms. Also called embryology.

dissipative structures   The second law of thermodynamics predicts that energy will go from an organized state containing potential energy to a disorganized state. The most efficient means of getting to the less organized state may be via an organized structure, like a tornado or the vortex of water as it drains out of a bathtub. These dissipative structures are also called self-organizing structures. Life itself is postulated to be a dissipative structure.

dominant   Describes a gene, or more correctly an allele, that allows for the appearance of a given trait in an organism that can obscure the expression of an alternative, so-called recessive gene/allele. If brown eye genes are dominant (which they are), a person with such genes will have brown eyes, even if the genes for other-colored eyes (blue eyes, say) are present.

electrophoresis   A process for assessing and comparing large, organic molecules. If soaked into a gel and then subjected to electrical charge, the molecules will migrate through the gel depending on their size and charge. Electrophoresis was used to prove that most organisms (including humans) are genetically more diverse than alike, and thus that species contained a great deal of variation that supported adaptation and evolution.

emergence   The principle that states that the properties of a system may not be completely predicted by the sum of the properties of its parts. Life is considered an emergent system, since, so far, it cannot be entirely explained by the properties of organic molecules that make it up.

entropy　The second law of thermodynamics, stating that the overall tendency of a system is toward greater randomness. Macroscopically, systems appear to go from states of disparate organization to states of greater uniformity (for example, solutions of different colors will blend to one color if allowed to mix). Microscopically, the end state is one of disorganized random distribution of particles (the molecules of all the solutions mix together, willy-nilly).

epigenetic　Referring to the successive specialization of previously undifferentiated cells into specialized organs and tissues during ontogeny.

essentialism　In biology, essentialism taught that elements of nature, including species, had an ideal and constant essence or quality that did not change—a concept traced back to Aristotle.

eugenics　Literally "good birth," coined by Darwin's cousin Frances Galton in 1864. Belief that careful control of human reproduction could result in the improvement of humankind, even to the point of genius becoming the "birthright of every person."

evolutionarily stable strategy　The principle, invented by J. B. S. Haldane, predicting stable physical and behavioral characteristics of organisms within their environments. An ESS is stable because it grants the organism optimal protection from forces of selection—predation, sexual selection, society, parasitism, etc.

gemmules   Particles of heredity postulated by Darwin, conceptual predecessors to the genes.

gene   The unit of inheritance. The molecular (DNA) code, or instructions, for the production of a protein molecule. The protein may be involved in the structure of a cell or may be involved in the regulation of other genes, switching them on or off.

genetic drift   The random accumulation or depletion of specific genes (alleles) in a population of organisms. Drift emphasizes the notion that genes (alleles) may be lost to or fixed into a population by chance, and not only because of selective forces.

gradualists   Evolutionary theorists who believe that evolutionary change of organisms happens slowly and by small increments. Darwin insisted that evolution happened gradually, and many of his contemporaries disagreed. A century and a half later, the argument still rages, and fires the split between such supernumaries as Richard Dawkins (gradualist) and the late Stephen Jay Gould.

heterozygous   When a diploid organism, one having two sets of chromosomes, has two different alleles (sets of DNA code) for the same gene.

homozygous   When a diploid organism, one having two sets of chromosomes, has the same allele (set of DNA code) for the same gene.

hybridization The crossing of two individuals of different species or varieties. Offspring of two different species (true hybrids) are usually sterile. Darwin tried hard to explain this phenomenon, and produced the least comprehensible section of *The Origin of Species* as a result.

invertebrates Animals without a backbone.

laissez-faire French for "leave it alone," policy of not regulating markets. In the tradition of Adam Smith's *Wealth of Nations* and also of social Darwinism, where failure and starvation of human populations are the natural means of balancing and managing populations and economies, rather than random cruelties that ought to be eliminated.

linear Describes a system in which the result of any external action is always proportional, and thus always precisely predictable.

macroevolution Change at the species level. Evolution that produces new species.

materialism The philosophy claiming that everything in the universe is produced out of matter and obedient to the observable laws of physics and chemistry. Leaves no room for a "spirit world."

memes Entities postulated by Richard Dawkins in his 1976 book, *The Selfish Gene,* to be the intellectual equivalent of genes, social constructions that are self-perpetuating and subject to selection for their survivability. Examples include bird songs and cultural institutions like cities and religions.

microevolution   A change in organisms below the species level, producing physical change within organisms of a species.

mutable   Changeable, as in species. Immutable means the opposite: unchangeable.

mutation   Initially used to mean rapid change exhibited by organisms within a single generation. Mutated organisms were often called sports or monsters. These changes could have been caused by nongenetic factors. We now restrict mutation to mean changes in the molecular nature of DNA. Mutations at this level may be neutral, meaning they may not result in measurable changes in an organism's outward appearance or functioning.

natural selection   The driving force behind change in organisms and species described by Darwin and Alfred Russel Wallace. The elements of natural selection are a population exhibiting variation—no two organisms are exactly alike—and an overabundance of organisms exploiting limited resources. The result is selection, or as Wallace put it "Why do some live and some die?" Only the organisms best suited for the environment and for competition for mates will reproduce, leaving their genetic material to persist.

New Synthesis   The re-evaluation and reconfirmation of Darwin's theory in the first half of the twentieth century. Several theorists from different disciplines of biology (including genetics, population biology, ecology, and paleontology) re-explained Darwinian evolution by natural selection using new discoveries.

ontogeny   The development of an organism from fertilization through embryonic growth to adulthood.

pangen   The material basis of heredity hypothesized by Hugo de Vries in 1889. Forerunner to the gene.

parsimonious   Describes the simplest scientific explanation for a phenomenon. It is more parsimonious to believe the earth revolves around the sun than to suggest epicycles to explain the movements of the sun and the planets around the earth.

parthenogenesis   Natural reproduction by cloning. Reproduction of offspring that are genetically identical to the (female) parent. Common in plants and also occurs rarely in lizards, always in rotifers, and rarely in birds.

phenotype   The physical expression of a genome; the observable shape and behavior of an organism as coded for by its particular set of genes.

phylogeny   Evolution of organic types, species, races, classes.

polyploidy   The condition of multiple copies of genomes—genes and chromosomes—in the cell of one organism. Most complex organisms are diploid, specifically having two copies, and some plants produce more than two copies under certain conditions.

post-normal science   Describes a soft system of scientific inquiry, where the facts are uncertain, social values of participants are considered, and complexity is acknowledged.

Principia   Written in 1687 by Sir Isaac Newton explaining the laws of physics and gravity. This work set the standard for scientific writing and was an intellectual progenitor of *The Origin of Species.*

punctuated equilibrium   The theory that species evolution (macroevolution) occurs in fits and starts. Postulated by Niles Eldredge and Stephen Jay Gould after they observed that fossils in a region remain the same for long periods of time before suddenly undergoing dramatic morphological change.

recessive   Genes (alleles) that are not expressed when in the presence of dominant genes (alleles). Recessive genes will be expressed only when they are homozygous, that is, when an organism has two of the same recessive allele.

saltationists   Evolutionary theorists who believe that species evolve suddenly, by jumps. Darwin did not believe that evolution proceeded this way, and instead insisted that "nature does not make jumps."

scala naturae   The organization of nature in a hierarchical, linear fashion, from the least perfect atom of matter to the epitome of perfection—man. Popular concept beginning with Aristotle (384–322 B.C.).

special creation   A theory of Darwin's time stating that every species was created specifically for its place and its time.

speciation   The development of new species from old species. In a word, the problem that Darwin set out to solve: How were new species derived from old species?

species   (An evolutionary biologist asked to define this will laugh.) The question of just what a species is has been debated and is not firmly resolved. Best bet: a reproductively isolated community occupying a specific niche in nature.

spontaneous generation   The generation of living forms from nonliving material. Includes notions from the historical belief that flies are produced in rotting meat to current theories encompassing the production of self-replicating molecules in nature from conditions of the early Earth.

survival of the fittest   The phrase coined by Herbert Spencer to describe Darwin's theory and accepted by Darwin only after much internal debate. Sets the tone for early interpretations that emphasized conflict and competition and made the theory unpopular with many religious and ethical thinkers.

sympatric speciation   The theory for species evolution due to altered reproduction by a small group within a larger population. If a few female peacocks, for example, suddenly decided they liked males with short tails, theoretically the population could split into two species. Never observed in nature.

theism   The belief that everything, in particular every species, was created in a special act by God.

theory   A framework of working principles used to test and establish explanations for related phenomena. Not the same as a hypothesis, or guess, in that a theory is in general accepted as the best current explanation for observed nature.

transmutation   To change into another nature, condition, form, or substance—word used by Darwin and contemporaries to designate change of one species into another.

tropism   The attraction exhibited by an organism. Plants exhibit tropism to the sun, turning their leaves to face it.

uniformitarianism   The theory of formation of the physical Earth based on mundane events—events experienced within human lifetimes, like earthquakes, volcanic eruptions, storms. The theoretical opposite to catastrophism.

vera causa   A literary device using a known and observed phenomenon to illustrate that a mechanism exists for a phenomenon in question. Newton used the movement of a pendulum as a vera causa to prove gravity; Darwin used the breeding of domestic animals to demonstrate artificial selection, and artificial selection was the vera causa he used to prove natural selection.

Wealth of Nations   The book by Adam Smith published in 1776 describing markets as self-regulating entities. Smith pointed out that individuals within the market make decisions to optimize their own income, and ultimately in turn optimize the market, making distribution, production, and pricing adjust to the demand of buyers. Smith also describes diversification of production, where individuals find their own niche for producing more and more specialized products to fit buyers' more specific needs and desires.

X-ray crystallography   T ohe use of reflected X rays to reveal the shape of molecules. Used by Watson and Crick to deduce the shape of DNA.

Bibliography

Bowler, Peter J. *Evolution: The History of an Idea*. Berkeley, Los Angeles, London: University of California Press, 1984.

Clark, Ronald W. *The Survival of Charles Darwin: A Biography of a Man and an Idea*. New York: Random House, 1984.

Darwin, Charles. *The Autobiography of Charles Darwin*. Edited by Frances Darwin. New York: Appleton & Company, 1893; New York: Amherst, 2000.

————. *The Origin of Species*. London: John Murray, 1872; London: Senate Studio Editions Ltd., Princess House, 1994.

————. *The Voyage of the Beagle*. London: Henry Colburn, 1839; London: Penguin Group, Penguin Books Ltd., 1989.

Dawkins, Richard. *Climbing Mount Improbable*. W. W. Norton, 1996.

————. *The Selfish Gene*. 2nd ed. Oxford: Oxford University Press, 1989.

Depew, David J., and Bruce H. Weber. *Darwinism Evolving*. Cambridge, MA, and London: A Bredford Book, The MIT Press, 1995.

De Waal, Frans. *The Ape and the Sushi Master: Cultural Reflections of a Primatologist*. New York: Basic Books, 2001.

Futuyma, Douglas J. *Evolutionary Biology*. 3rd ed. Sunderland, MA: Sinauer Associates, Inc., 1998.

Gayon, Jean. *Darwinism's Struggle for Survival.* Translated by Matthew Cobb. Universite Paris 7-Denis Diderot 1992: Cambridge University Press, 1998.

Glick, Thomas F. *The Comparative Reception of Darwinism.* Austin and London: University of Texas Press, 1972.

Goodwin, Brian. *How the Leopard Changed Its Spots: The Evolution of Complexity.* Princeton, NJ: Princeton University Press, 1994.

Gould, Stephen Jay. *Ever Since Darwin: Reflections in Natural History.* New York, London: W. W. Norton, 1977.

———. *The Panda's Thumb: More Reflections in Natural History.* New York, London: W. W. Norton, 1980.

———. *Wonderful Life.* New York, London: W. W. Norton, 1989.

Kauffman, Stuart. *At Home in the Universe: The Search for Laws of Self-Organization and Complexity.* New York, Oxford: Oxford University Press, 1995.

Larsen, Edward J. *Summer for the Gods: The Scopes Trial and America's Continuing Debate over Science and Religion.* Cambridge, MA: Harvard University Press, 1998.

Mayr, Ernst. *The Growth of Biological Thought.* Cambridge, MA, and London: The Belknap Press of Harvard University Press, 1982.

———. *One Long Argument: Charles Darwin and the Genesis of Modern Evolutionary Thought.* Cambridge, MA: Harvard University Press, 1991.

———. *What Evolution Is.* New York: Basic Books, 2001.

Morris, Richard. *The Evolutionists: The Struggle for Darwin's Soul.* New York: W. H. Freeman, 2001.

Quammen, David. *The Song of the Dodo.* New York: Scribner, 1996.

Raby, Peter. *Alfred Russel Wallace: A Life*. Princeton, NJ: Princeton University Press, 2001.

Ridley, Matt. *The Red Queen: Sex and the Evolution of Human Nature*. London: Penguin Books, 1993.

Smith, John Maynard. *Did Darwin Get It Right?* New York: Chapman and Hall, 1988.

Sterelny, Kim. *Dawkins vs. Gould: Survival of the Fittest*. Cambridge, UK: Icon Books, 2001.

Strick, James E. *Sparks of Life: Darwinism and the Victorian Debates Over Spontaneous Generation*. Cambridge, MA, and London: Harvard University Press, 2000.

Weiner, Jonathan. *The Beak of the Finch*. New York: Vintage Books, 1994.

Wilson, Edward O. *Consilience: The Unity of Knowledge*. New York: Alfred A. Knopf, 1998.

——— . *Sociobiology: The New Synthesis*. Cambridge, MA, and London: Belknap Press of Harvard University Press, 1975 and 2000.

Index

acquired characteristics, 16–17,
 89, 122, 125n, 134–35, 151
adaptation, 11, 16, 59, 141, 154,
 178, 179–81, 191
 arguments against, 164
 gene selection and, 168–74, 179
 luck vs., 180
 See also natural selection
adaptive landscapes, 120
agnostic, 143, 191
agriculture, 115, 140, 150, 151,
 152
alkaptonuria, 114
allele, 117, 191, 197
 reproduction of, 120, 173–74
 See also gene
allopatric speciation, 176–77,
 191–92
altruism, 141, 164–66, 168–70
Anglican church, 21, 22, 23,
 26–27, 31, 41, 77–78
animalcules, 7, 8, 192
archaeopteryx, 80, 123, 124, 192
Aristotle, 5, 8, 10, 11, 90, 136,
 196, 201
artificial selection, 62, 64–65, 127,
 203
atheism, 86, 125n, 192
attractors, 185, 187, 188, 192

Baer, Karl Ernst von, 90–92, 137
balancing selection, 177
Bastian, Henry, 93–94

Bates, Henry Walter, 45, 48–50
Bateson, William, 95–96, 97–98,
 108–16, 118, 175
Beagle voyage, 29–36, 62n, 66, 71,
 73
Beloussov, Lev, 182
Bible, 6, 21, 96, 193
 creation science and, 148–49
 literal reading of, 143–45, 146,
 149, 194
 metaphorical reading of, 28, 82
biblical criticism, 146
biogenetic law (Haeckel), 92,
 136–37
biometricians, 88, 96, 109, 117,
 118, 135, 192
 game theory and, 164–68
 See also mathematics
birth control, 55, 71
black peppered moth, 121–23
Boltzmann, Ludwig, 155, 158
Boole, George, 189, 192–93
Boolean networks, 187–89,
 192–93
botany. *See* plants
brain, 79–80
breeding. *See* genetics and
 heredity
Bridges, Calvin, 113n
British Museum, 49, 74
Bruguiere, Jean Guillaume, 15
Bryan, William Jennings, 142–47
Buffon, 9–13, 14, 28

209

Carnot, Sadi, 158, 159
catastrophism, 28, 32–34, 193, 203
Catholicism, 98, 138, 140
cells, 88–89, 196
    division of, 89, 113–14
Chaga's disease, 62n
chain of being. *See* scala naturae
Chambers, Robert, 50, 71
chance. *See* randomness
chaos, 152, 184, 185–86, 188, 189,
    193, 194
Chetverikov, Sergei, 126–27
chromosomes, 113–14, 116,
    124–25, 193
    first identification of, 88–89
classification, 10–11, 14–15, 53
Clausius, Rudolf, 158
cloning, 172n, 200
common origin, 65, 91, 193
communism, 125n, 138, 150–51
community. *See* group selection
comparative anatomy, 11, 41, 65,
    79
competition, 25, 59, 60, 131–34,
    141, 199, 202
    animal altruism vs., 164–65
complexity, 15, 61, 92, 133,
    183–86, 189, 193–94
computers, 184–85, 187
constrained development, 180–82
continuous modification, 133
cooperation, 141, 164–65, 168–70
Cope, Edward Drinker, 85
crab evolution study, 108–9
creationism, 6, 27, 142–49, 152, 194
    Darwinism vs., 3, 4, 70, 76
    Lamarckism and, 15–16
    special creation and, 70, 81,
        149, 201
    theism and, 8, 82, 202
    Wallace linked with, 44, 53

creation science, 147–49, 152
Crick, Francis, 160–62, 203
Cuvier, Georges, 9, 28

Darling, Cyril, 116
Darrow, Clarence, 142–46
Darwin, Charles, 1–4, 20–41, 43,
    56
    allies of, 72–81
    atheism of, 86, 125n
    background of, 20–24, 30, 46,
        48, 135
    Beagle voyage and, 29–36, 37,
        38, 62n, 66, 71
    caution of, 24, 27, 31–32, 39,
        40, 44, 69–70, 71, 80, 87
    common origin theory of, 65,
        91, 193
    critics of, 1–3, 61–62, 65,
        75–76, 79, 80–85
    developmental biology and, 90,
        91
    eugenics and, 3, 135–36
    evolution notebooks, 38–41
    evolution theory, 43, 56–67.
        See also natural selection
    Galton and, 2, 3, 84, 135
    gemmules postulate, 65, 84,
        114, 197
    gradualism and, 3, 94–95, 106,
        108–9, 116, 124, 174, 197
    human evolution and, 67,
        69–70, 82–83
    on hybridization, 198
    illnesses of, 57, 62n, 86
    influence of, 67, 87–92, 125n,
        126, 151, 155
    influences on, 24–25, 55, 115n
    legacy of, 155, 189
    personal life of, 56–57, 62n, 86,
        136

plant studies by, 83, 85
public image of, 71–72
on saltationist approach, 201
scientific methods of, 63–64
scientific reputation of, 37–38
social Darwinism confused
    with, 131, 142, 146
Soviet genetics and, 125n, 126,
    151
Spencer and, 132, 133
"survival of the fittest" phrase
    and, 131, 202
transmutation and, 38, 39, 203
vera causa of, 62, 203
Wallace and, 44–46, 53, 56–58,
    60–61, 82
writings by, 37, 38, 83, 85. *See
    also Origin of Species, The*
Darwin, Emma Wedgwood (wife),
    86, 136
Darwin, Erasmus (brother), 37, 71
Darwin, Erasmus (grandfather),
    20–22, 24, 30, 31, 38, 48, 135
Darwin, George (son), 149
Darwin, Leonard (son), 118
Darwin, Robert (father), 22, 23,
    24, 25–26, 29, 36, 86
Davenport, Charles, 139
Dawkins, Richard, 168–70, 172,
    183, 197, 198
deism, 8, 16, 194
demes, 119–20, 156, 173–74, 194
deoxyribonucleic acid (DNA), 65,
    152, 155, 193, 197, 199
  genome and, 169
  matching, 166
  structure of, 161–62, 194, 203
Depew, David, 189
developmental biology, 65, 90–92,
    111, 112, 136–37, 178–80,
    186–89, 195

diploid organisms, 197, 200
discontinuous evolution. *See*
    saltationists
dissipative structures, 184–86, 195
DNA. *See* deoxyribonucleic acid
Dobzhansky, Theodosius, 123,
    126, 127, 171, 177
dominant, 104, 121, 195
double helix, 162, 194
dynamic systems, 184–85

earth, age/formation of, 28,
    32–34, 80–81
earthquake, 33–34
earthworms, Darwin study of, 85
ecology, 136, 182–84
effective populations. *See* demes
Einstein, Albert, 2, 158, 189
Eldredge, Niles, 149, 174–76, 178,
    201
electrophoresis, 127–28, 171, 195
embryology. *See* developmental
    biology
emergence, 157, 195–96
emotional expression, 85
energy. *See* thermodynamics
entropy, 158–60, 184–86, 196
epigenetic, 179, 196
equilibrium, 160, 177, 185, 188–89
essentialism, 10, 12, 15, 19, 196
eugenics, 3, 88, 127, 134–41, 171,
    172, 196
evolutionary stable strategy,
    166–68, 196–97
evolution theory, 1–4, 24, 56–67,
    87–96
  backlash against, 142–49,
    170–74
  censorship of, 76–77, 142–49
  co-options of, 3, 92, 95–96,
    131–41

evolution theory *(continued)*
different direction of, 111–12
early versions of, 12–17, 24, 38–41, 39, 50, 51
impact of, 131–62
influences on, 10–11, 54–56
Lamarck and, 13–17, 74, 76, 84, 132
New Synthesis, 123–27, 174, 175, 179, 199
pace and, 94–96. *See also* gradualists; punctuated equilibrium; saltationists
resistance to, 1–3. *See also* creationism
social agendas and, 40, 71, 79, 125n
sociobiology and, 168–70
Spencer and, 39, 75, 131–34
"survival of the fittest" phrase and, 131–34, 202
theoretical biology and, 188–89
tree branch model, 53, 66, 74, 91, 132
Wallace and, 43–46, 53–60, 82
See also adaptation; Darwin, Charles; natural selection
extinction, 15–16, 59, 66, 167, 173, 181

fertilization, 100, 105n
Fibonacci equations, 181–82
Fisher, Ronald A., 104n, 117–21, 123, 125, 126, 156, 171
Fitzroy, Robert, 29–36, 38
Flemming, Walther, 88–89
Ford, E. B., 122
fossils, 8, 10, 13, 15, 28, 148, 174–79
Darwin and, 31, 34–35, 67, 80, 123, 192

punctuated equilibrium and, 175–76, 201
Founder Effect, 176
Fox, William Darwin, 26
Franklin, Rosalind, 161, 162
fruit flies, 112–13, 125, 126, 171, 172, 179
fundamentalists, 142–49

Galápagos Islands, 35–36, 38, 63, 64, 66, 153–54
Galton, Francis, 2, 3, 84, 87–88, 95, 96, 105, 109, 117, 135, 196
game theory, 152, 164–68
Garrod, Archibald, 114–15
gemmules, 65, 84, 114, 197
gene, 163, 179–82, 197
allele and, 191
Boolean model, 187–89, 193
as Dawkins abstraction, 170
DNA and, 152, 193, 194, 197
dominant, 104, 121, 195
earlier concepts of, 65, 84, 114, 197, 200
interactions, 128–29, 173
misunderstandings of, 140–51
naming of, 114
natural selection and, 172–73, 177, 199
new species creation and, 178
pangen as forerunner of, 200
recessive, 104, 114, 117–18, 201
trait survival and, 168–70
variation effects, 117–18
See also chromosomes; demes; genome/genotype; mutation
gene selection, 168–74, 179
gene sequencing, 128
genetic assimilation, 179
genetic determinism, 171, 172

genetic diseases, 114–15
genetic diversity, 195
genetic drift, 120, 164, 197
genetics and heredity, 60, 111,
   112–29, 174
  acquired characteristics theory,
    16–17, 89, 122, 125n, 134–35,
    151
  advanced findings, 163
  agricultural applications,
    115–16
  Darwin's studies, 40, 62–65,
    83–85, 86, 105n
  DNA and, 65, 155, 161, 162,
    194
  eugenics and, 139–40
  Galton studies, 87–88, 105, 117,
    135
  hybridization, 10, 13, 115, 198
  Lamarckian, 16–17, 89
  Mendel and, 98, 100–108, 114,
    117, 135, 151
  Mendelian ratio, 103–4, 112
  Morgan and, 112–13
  natural selection and, 123,
    124–25, 165–66
  paleontology and, 123
  pangen and, 114, 200
  as particulate, 105–6, 108
  questions in 1900, 97–98
  recessive vs. dominant traits,
    104–5, 114, 117–18, 195, 201
  Soviet, 125n, 126–27, 150–52
  stabilizing selection and,
    177–78
  variation and, 127–28
  See also gene; population
    genetics
genome/genotype, 155, 165, 172,
   174, 177, 200
  explanation of, 169

geology, 23, 27–28, 32–34, 38, 39,
   58, 81–82
germ theory of disease, 93, 94
Goldschmidt, Richard, 124, 178–79
Goodwin, Brian, 181–82
Gore, Charles, 96
Gosse, Philip, 81n
Gould, John, 37, 38
Gould, Stephen Jay, 149, 174–76,
   178–83, 197, 201
gradualists, 3, 4, 85, 106, 108–9,
   111, 116, 123, 197
  challenges to, 94–95, 124, 149,
    174–76
Grant, Robert, 24, 39
Gray, Asa, 57, 78
group selection, 164–65, 173–74

Haeckel, Ernst, 65, 92, 136–37
Haldane, J. B. S., 120–21, 122,
   123, 166, 196–97
Hamilton, W. D., 165–66
Hardy-Weinberg equation, 117
Harris, H., 128
Heisenberg's Uncertainty
   Principle, 119, 155–56
Henslow, John, 27–30, 77
heredity. *See* genetics and heredity
Herschel, John, 27, 28, 62
heterozygous, 126, 128, 171, 197
hierarchy of nature. *See* scala
   naturae
hierarchy of selection, 182–83
Hitler, Adolf, 92, 137
Holocaust, 3, 140
homozygous, 125–27, 140–41,
   173, 174, 197
Hooker, Joseph, 40–41, 44, 56,
   57, 58, 73, 75
"hopeful monsters," 178
Hubby, J. L., 128

Humboldt, Alexander von, 27, 28, 49

Hume, David, 25

Huxley, Julian (grandson), 123, 124

Huxley, Thomas, 2, 41, 44, 72–80, 82, 87, 93–95, 98, 123, 132, 191

hybridization, 10, 13, 115, 198

industrial melanism, 121–23

Inherit the Wind (play/film), 146, 147

intelligence, 135, 138, 168

intelligence tests, 88

intelligent design, 4

invertebrates, 14–15, 24, 26, 109, 198

Jews/Judaism, 136, 137, 140, 147

Jukes and Kallikaks, 139–40

Kauffmann, Stuart, 186–89, 192–93

Kettlewell, H. B. D., 122

Kimura, Motoo, 164

kin selection, 165–66, 168–70

Kropotkin, Peter, 141

laissez-faire, 55, 71, 133–34, 198

Lamarck, Jean Baptiste, 9, 13–17, 19, 20, 24, 71, 74, 76, 84, 89, 122, 125n, 132, 134

Leeuwenhoek, Anton van, 7, 192

Lenin, V. I., 115, 150

Lewontin, Richard, 128, 171–72, 173, 179, 180–81

linear, 198
    nonlinear systems vs., 184, 189, 193
    Spencer's theory as, 132–34

Von Baer's and Haeckel's theories as, 91–92
    See also complexity; scala naturae

Linnaeus, Carolus, 10–11, 14

Linnean Society, 57, 58, 70, 82

luck, 180

Lyell, Charles, 32–33, 37–38, 39, 43–44, 53, 73, 76
    Darwinism and, 56–57, 58
    refutation of Lamarck by, 132

Lysenko, Trofim, 125n, 150–52

macroevolution, 178, 198, 201

male chromosome, 113n

male sperm, 100, 105

Malthus, Thomas, 25, 37, 54–56, 71, 115

Martineau, Harriet, 37, 39, 71

Marx, Karl, 87

materialism, 198

mathematics, 25, 61, 181–82
    Boolean networks, 187–89
    evolution models, 117–18, 121
    Galton and, 2, 87–88, 95, 96
    Lysenko's rejection of, 151
    Mendel and, 102, 103, 105
    Weldon and Pearson and, 109–11
    See also biometricians; game theory; probability

Mayr, Ernst, 2, 123, 124, 172–73, 176–77, 191

McCormick, Robert, 29–30

meiosis, 113–14

memes, 170, 198

Mencken, H. L., 143

Mendel, Gregor, 96, 98–108, 110, 112, 114, 117, 118, 135, 151

metabolism, 159

microevolution, 178, 199

microorganisms, 94, 192, 193
microscope, 7, 88–89, 192
mitosis, 89
Mivart, St. George Jackson, 82–83
molecular biology, 156–57, 163,
    171, 174
Monkey Trial (1925), 142–47
moral philosophy, 24–25
Morgan, Thomas Hunt, 112, 116,
    125, 138
moths, light vs. dark, 121–22
Muller, Herman J., 124–25, 126,
    127, 138, 141, 150, 152, 161
mutable, 10, 50, 70, 199
mutation, 112–18, 124–26, 176,
    179, 199
  chance and, 164
  fuzzy concept of, 116, 121–22
  sudden change and, 95
  variation vs., 106

Nageli, Karl Wilhelm von, 107
natural history collections, 6–9,
    11
  Darwin and, 26–38
  Huxley and, 73–74
  Wallace and, 43–45, 48–54, 58
natural selection, 58–67, 100
  accessibility of theory, 2–3
  altruism in, 141, 164–66
  arguments against, 112–16,
    164, 182, 188
  chaos and, 152, 189
  comeback of theory, 117–27
  conflict implications, 134, 141
  Darwin on, 62–67, 83, 203
  Darwin's and Wallace's separate
    formulation of, 58, 199
  description of, 59–60, 199
  developmental biology and,
    187–89

first natural proof of, 122
game theory and, 152, 164–68
genes and, 172–73, 177, 199
group selection and, 164–65,
    173–74
hierarchy of selection and,
    182–83
industrial melanism and,
    121–22
inherited characteristics and,
    83–85, 118
Mendelian heredity and, 100,
    104–6, 108, 117
New Synthesis, 123–27, 199
reaction to, 69–96
sociobiology and, 168
Spencer's co-option of, 133
stabilizing selection and,
    177–78
  See also variation
natural theology, 27, 49, 159, 180
nature, 6, 20, 26–27, 48, 186
  patterns in, 181–82
nature vs. nurture, 171
Nazis, 3, 137, 138, 140, 150
neo-Darwinism, 151, 175
neptunism, 28
neutrality theory, 164
New Synthesis, 123–27, 174, 175,
    179, 199
Newton, Isaac, 2, 9, 10, 25, 31, 58,
    60, 155, 189, 201, 203
Noah's flood, 7, 28, 148
nonlinear systems, 184, 189, 193.
  See also chaos; complexity;
    randomness

one-celled organisms, 119, 192
ontogeny, 196, 200
"ontogeny recapitulates
    phylogeny," 65, 92, 136

Origin of Species, The (Darwin), 39,
    75–81, 94, 98, 137, 198, 201
  clarity of, 2, 62
  as Darwinism defense, 62–67,
    155, 183
  on fossil record gaps, 123
  publication of, 57–58
  revisions of, 67, 83
  sales of, 70
  "survival of the fittest" phrase
    and, 131
Overton, William R., 148
Owen, Richard, 37, 41, 65, 75–76,
    79–80

paleontology, 111, 123, 149,
    174–76, 178
Paley, William, 27, 180
pangen, 114, 200
parsimonious, 200
parthenogenesis, 200
particulate inheritance, 105–6,
    108
Pasteur, Louis, 94
Pauling, Linus, 161–62
Pearson, Karl, 108, 110, 111, 117,
    118, 137
phenotype, 129, 177, 178, 179,
    200
phylogeny, 65, 92, 136, 200
pigeon breeding, 40, 63–64
plants
  Darwin's studies, 83, 85
  Lamarck's studies, 13–14
  leaf/flower petal distribution,
    181–82
  Mendel's experiments, 99–102
  mutation vs. new varieties, 95,
    116
  Unger's evolution theory of, 99
pollution, 108–9, 121–23

polyploidy, 116, 200
population genetics, 121, 126–27,
    189
  allopatric speciation and,
    176–77
  altruism and, 164–65
  demes and, 119–20, 156, 173
  ESS and, 166–68, 196–97
population theory, 54–56, 71,
    115, 198
post-normal science, 154–57, 184,
    200
predictability, 184–86, 188, 189
Priestley, Joseph, 22
Prigogine, Ilya, 186
Principia (Newton), 25, 201
probability, 60–61, 120, 155, 184,
    189
punctuated equilibrium, 149,
    174–76, 178, 182, 201

quantum mechanics, 119, 155–56,
    159, 161, 189

racism, 92, 136, 137, 138, 140,
    141, 146
randomness, 141, 155–56, 164,
    189
  entropy and, 159, 196
  inherited traits, 105, 119–20
Rappleyea, George, 142
rate genes, 178
recessive, 104–5, 114, 117–18,
    195, 201
Red Queen hypothesis, 177–78
reductionism, 156, 168, 180, 183
regulatory genes, 193
religion. *See* Anglican church;
    Bible; creationism; theism
reproduction, 200, 202
  altruistic, 165–66

demes and, 119–20, 156, 173, 194

eugenic control of, 137, 138, 140, 196

group selection and, 173–74

natural selection and, 59, 60, 63, 64–65, 199

"selfish gene" and, 168–70

Romanes, Georges, 107

Russia. *See* Soviet Union

St. Hilaire, Geoffrey, 9

saltationists, 94–95, 106, 108, 109, 110, 111, 121, 175, 201

scala naturae, 11, 15, 74, 90, 92, 132, 136, 201

Schrödinger, Erwin, 159–60, 161

science education, 147

scientific career, 48–49, 72, 98

Huxley and, 41, 73–74, 79

Scopes, John, 142–47

Sedgwick, Adam, 27–28, 32, 33, 39, 76

Selfish Gene, The (Dawkins), 168–70, 198

self-organizing structures, 184–86, 195

sex-linked traits, 113n, 139

Simpson, George Gaylord, 123

Smith, Adam, 24–25, 55, 198, 203

Smith, John Maynard, 166–68, 189

social Darwinism, 3, 131, 133–41, 142, 146, 171, 198. *See also* eugenics

social insects, 164, 165, 166

sociobiology, 168–74

Soviet Union, 115, 125, 138, 147

Lysenkoism and, 150–52

spandrels, 180

special creation, 70, 81, 83, 149, 201

speciation, 92, 124, 153–54, 201

allopatric, 176–77, 191–92

Darwin's ideas on, 178n

Lamarck and, 13–17, 20

macroevolution and, 177, 178, 198

punctuated equilibrium and, 149

sympatric, 176n, 202

species, 202

hybridization, 10, 13, 115, 198

intermediate forms, 80, 192

microevolution and, 199

varieties vs., 116

See also adaptation; mutation; natural selection; variation

Spencer, Herbert, 39, 73, 75, 131–34, 202

spontaneous generation, 7–8, 92–94, 202

stabilizing selection, 177–78

Stalin, Joseph, 125n, 150, 151

statistics. *See* biometricians; mathematics

Stebbing, T. R., 111

sterility, 13, 198

sterilization, 137, 138, 140

"survival of the fittest," 59, 60, 131–34, 141, 202

sympatric speciation, 176n, 202

theism, 8, 82, 202

theocracy, 21, 48–49, 72, 79

theory, 202

thermodynamics, 157–60, 184, 185, 186, 189, 195, 196

Thomson, William, 80–81

transformationism, 90–91

transmutation, 38, 39, 203

tropism, 119, 203

Tyndall, John, 93–94

uncertainty, 119, 155–56. *See also* chaos; complexity; randomness
Unger, Franz, 99
uniformitarianism, 32–33, 34, 58, 95, 203
Unitarianism, 22, 86
Ussher, James, 28

Van Valen, Leigh, 177–78
variation
environment and, 11, 16, 154, 172
gene type and, 117–18, 127
heterozygosity and, 126, 128, 171, 197
mechanism for, 106, 108, 117
natural selection and, 63–64, 83, 85, 100, 118, 121, 199
wild populations and, 127
Vavilov, Nikolai, 115, 125, 150–52
vera causa, 62–63, 203
Vestiges of the Natural History of Creation (Chambers), 39, 50, 71
Vries, Hugo de, 95, 110, 112, 114, 116, 200
vulcanism, 28

Waddington, Conrad Hal, 179
Wallace, Alfred Russel, 1, 39, 72, 87
evolution theory of, 43–46, 53–60, 61–62, 70, 82, 182, 199

family background, 46–47, 73
Wallace, Herbert, 51
Wallace, John, 49
Wallace, William, 47, 48, 49
Watson, James D., 161, 203
Wealth of Nations (Smith), 24–25, 198, 203
Weber, Bruce, 189
Wedgwood, Josiah, 21–22, 24, 29, 30, 31, 48
Weismann, Leopold, 89–90, 135
Weldon, Walter, 95–96, 108–11
Wilberforce, Samuel, 77–78
Wilkins, Maurice, 161
Williams, George C., 165
Wilson, E. O., 4, 168, 170, 172
working class, 40, 47, 63–64, 78–79, 79
Wright, Sewall, 118–21, 123, 125, 126, 156, 171, 179
Wynne-Edwards, E. C., 164–65

X chromosome, 113n
X-ray crystallography, 161, 162, 203
X-ray mutations, 125

Zhabotinsky, Anatol, 182
Zoonomia (C. Darwin), 38–39, 43
Zoonomia (E. Darwin), 21, 22, 24, 38

A NOTE ON N
AND TER

In 1779 there was no state of Maine
province of Massachusetts. Some
changed. Majabigwaduce is now c
is Bucks Harbor, and Falmouth is
plantation (properly Plantation Nu
Orphan Island is Verona Island, Lon
River) is now Islesboro Island, W
Cape Jellison and Cross Island is to

The novel frequently refers to 's
'schooners'. They are all, of cou
that they are all boats, but pr
square-rigged, three-masted ves
the USS *Constitution*) or a ship o
Nowadays we think of a sloop
but in 1779 it denoted a three-m
smaller than a ship and distingu
deck (thus no raised poop de
square rigged (meaning they (
from crosswise yards). A brig,

square-ri
Schoone
with fo
centre
variat'
there

Mos
nam
F (w
the n
(with

Excerpt of letter from the Massachusetts Council, to Brigadier-General Solomon Lovell, July 2nd, 1779:

> *You will in all your operations consult with the Commander of the fleet that the Naval Force may cooperate with the troops under your command in Endeavoring to Captivate Kill or Destroy the whole force of the Enemy there both by sea & land. And as there is good reason to believe that some of the Principal men at Majorbagaduce requested the enemy to come there and take possession you will be peculiarly careful not to let any of them escape, but to secure them for their evil doings . . . We now commend you to the Supream being Sincerely praying him to preserve you and the Forces under your Command in health and safety, & Return you Crowned with Victory and Laurels.*

From a postscript to Doctor John Calef's Journal, 1780, concerning Majabigwaduce:

> *To this new country, the Loyalists resort with their families . . . and find asylum from the tyranny of Congress, and their taxgatherers . . . and there they continue in full hope, and*

*pleasant expectation, that they may soon re-enjoy the liberties
and privileges which would be best secured to them by
the . . . British Constitution.*

Letter from Captain Henry Mowat, Royal Navy, to Jonathan
Buck, written aboard HMS *Albany*, Penobscot River, June
15th, 1779:

*Sir, Understanding that you are at the head of a Regiment of
the King's deluded Subjects on this River and parts adjacent
and that you hold a Colonel's Commission under the influence
of a body of men termed the General Congress of the United
States of America, it therefore becomes my duty to require you
to appear without loss of time before General McLean and the
commanding Officer of the King's Ships now on board the
Blonde off of Majorbigwaduce with a Muster Roll of the People
under your direction.*

ONE

There was not much wind so the ships headed sluggishly upriver. There were ten of them, five warships escorting five transports, and the flooding tide did more to carry them northwards than the fitful breeze. The rain had stopped, but the clouds were low, grey and direful. Water dripped monotonously from sails and rigging.

There was little to see from the ships, though all their gunwales were crowded with men staring at the river's banks that widened into a great inland lake. The hills about the lake were low and covered with trees, while the shoreline was intricate with creeks, headlands, wooded islands and small, stony beaches. Here and there among the trees were cleared spaces where logs were piled or perhaps a wooden cabin stood beside a small cornfield. Smoke rose from those clearings and some men aboard the ships wondered if the distant fires were signals to warn the country of the fleet's arrival. The only people they saw were a man and a boy fishing from a small open boat. The boy, who was named William Hutchings, waved excitedly at the ships, but his uncle spat. 'There come the devils,' he said.

The devils were mostly silent. On board the largest warship, a 32-gun frigate named *Blonde*, a devil in a blue coat and an oilskin-covered cocked hat lowered his telescope. He frowned thoughtfully at the dark, silent woods past which his ship slid. 'To my mind,' he said, 'it looks like Scotland.'

'Aye, it does,' his companion, a red-coated devil, answered cautiously, 'a resemblance, certainly.'

'More wooded than Scotland, though?'

'A deal more wooded,' the second man said.

'But like the west coast of Scotland, wouldn't you say?'

'Not unlike,' the second devil agreed. He was sixty-two years old, quite short, and had a shrewd, weathered face. It was a kindly face with small, bright blue eyes. He had been a soldier for over forty years and in that time had endured a score of hard-fought battles that had left him with a near-useless right arm, a slight limp, and a tolerant view of sinful mankind. His name was Francis McLean and he was a Brigadier-General, a Scotsman, commanding officer of His Majesty's 82nd regiment of foot, Governor of Halifax, and now, at least according to the dictates of the King of England, the ruler of everything he surveyed from the *Blonde*'s quarterdeck. He had been aboard the frigate for thirteen days, the time it had taken to sail from Halifax in Nova Scotia, and he felt a twinge of worry that the length of the voyage might prove unlucky. He wondered if it might have been better to have made it in fourteen days and surreptitiously touched the wood of the rail. A burnt wreck lay on the eastern shore. It had once been a substantial ship capable of crossing an ocean, but now it was a ribcage of charred wood half inundated by the flooding tide that carried the *Blonde* upriver. 'So how far are we now from the open sea?' he asked the blue-uniformed captain of the *Blonde*.

'Twenty-six nautical miles,' Captain Andrew Barkley answered briskly, 'and there,' he pointed over the starboard